CREATING SPATIAL INFORMATION INFRASTRUCTURES

Towards the Spatial Semantic Web

CREATING SPATIAL INFORMATION INFRASTRUCTURES

Towards the Spatial Semantic Web

Edited by
Peter van Oosterom • Sisi Zlatanova

CRC Press
Taylor & Francis Group
Boca Raton   London   New York

CRC Press is an imprint of the
Taylor & Francis Group, an **informa** business

Axel Smits is the creator of the figure in the center of the cover.

CRC Press
Taylor & Francis Group
6000 Broken Sound Parkway NW, Suite 300
Boca Raton, FL 33487-2742

© 2008 by Taylor & Francis Group, LLC
CRC Press is an imprint of Taylor & Francis Group, an Informa business

Library of Congress Cataloging-in-Publication Data

Creating spatial information infrastructures : towards the spatial Semantic Web /
    editors, Peter van Oosterom, Sisi Zlatanova.
        p. cm.
    Includes bibliographical references and index.
    ISBN 978-1-4200-7068-2 (hardback : alk. paper)
        1. Spatial data infrastructures. 2. Semantic Web. I. Oosterom, Petrus Johannes
    Maria van II. Zlatanova, Siyka. III. Title.

G70.212.C74 2008
025.04--dc22   2008008412

Visit the Taylor & Francis Web site at
http://www.taylorandfrancis.com

and the CRC Press Web site at
http://www.crcpress.com

Table of Contents

Chapter 7

Chapter 8

Chapter 9

Chapter 10

Chapter 11

Preface

The concept of information infrastructures is an important one in many countries and organizations for improving the management of distributed information. Many national and international efforts are ongoing to create such information infrastructures, in parallel with research activities to find answers to the remaining problems when realizing these information infrastructures; for example, what does the information offered by different organizations, within the same infrastructure, actually mean (the semantic aspect)? The focus of this book is on spatial information within a larger information infrastructure. The Introduction of this book will give a general overview of spatial information infrastructures and introduce the other chapters in the book.

This book has its roots in a Bentley BE research seminar with the same title: "Creating Spatial Information Infrastructures: Towards the Spatial Semantic Web," which was held June 11, 2007, in London. About half of the chapters in this book are related to a presentation at the seminar, and in all cases this was followed by a discussion with the audience. The other half of the chapters are from contributors invited by the editors in order to have a more complete coverage of the subject. Full chapters were reviewed and commented on by at least three other contributors to the book.

The above-mentioned seminar was the fourth in a series of BE research seminars. There are clear links with the previous BE research seminar topics (2004, 2005, 2006), which is obvious from the seminar titles:

- May 23, 2004, Orlando, Florida: Large-Scale 3D Geo-Information, the Integration of CAD and GIS (published as *Large-scale 3D Data Integration: Challenges and Opportunities*, edited by S. Zlatanova and D. Prosperi. ISBN 0-8493-9898-3, CRC Press, Taylor & Francis Group, Boca Raton, Florida, 2006).
- May 8, 2005, Baltimore, Maryland: Sustainable, High Accuracy, Infrastructure Information Management and Interoperability—A Framework for Homeland Security. (Selected papers were included in *Geospatial Information Technology for Emergency Response*, edited by S. Zlatanova and J. Li. Taylor & Francis Group, London, UK, 381.)
- June 11, 2006, Prague, Czech Republic: Advancing GIS for Infrastructure.

The editors would like to thank Bentley Systems for creating the unique opportunity to exchange research ideas on cutting-edge topics in the geospatial information and communication technology (geo-ICT) domain. The editors would further like to thank the chapter authors for their efforts, often during very busy times, to write chapters and improve their draft chapters based on the reviews provided. We do expect and hope that this book will provide background knowledge for those who are involved in realizing parts of the current spatial information infrastructure either in practice or in research.

Peter van Oosterom
Sisi Zlatanova
Delft, The Netherlands
October 2007

Editors

Peter van Oosterom is full professor and head of the GIS Technology Section, OTB Research Institute and Faculty of Technology, Policy, and Management, Delft University of Technology, The Netherlands. In 1990 he obtained a Ph.D. degree from Leiden University based on the thesis "Reactive Data Structures for GIS." His research interests include spatial databases; GIS architectures; spatial analysis; generalization, querying and presentation; Internet/interoperable GIS and Cadastral applications.

Sisi Zlatanova is an associate professor at the GIS Technology Section, OTB Research Institute for Housing, Urban and Mobility Studies, Delft University of Technology, The Netherlands. She received her Ph.D. degree from Graz University of Technology, Austria. Her research interests focus on three-dimensional aspects of geo-information and disaster management. She leads a theme group, "Geo-information for Crisis Management," within the GIS Technology Section.

Contributors

Henri J. G. L. Aalders
Delft University of Technology
Delft, The Netherlands

Alessandro Annoni
European Commission—Joint
    Research Centre
Ispra, Italy

Styli Camateros
Bentley Systems
Noida, India

Oscar Custers
Bentley Systems
Noida, India

Catherine Dolbear
Ordnance Survey
Southampton, United Kingdom

Anders-Friis-Christensen
European Commission—Joint
    Research Centre
Ispra, Italy

Chris Goad
Palatial, Inc.
Astoria, Oregon

Arco Groothedde
ITC/Kadaster
Enschede, The Netherlands

Thorben Hansen
National Survey and Cadastre
Copenhagen, Denmark

Frank van Harmelen
Vrije Universiteit
Amsterdam, The Netherlands

Glen Hart
Ordnance Survey
Southampton, United Kingdom

John R. Herring
Oracle Corporation
Nashua, New Hampshire

Ravikanth V. Kothuri
Oracle Corporation
Nashua, New Hampshire

Christiaan Lemmen
ITC/Kadaster
Enschede, The Netherlands

Rob Lemmens
ITC/Kadaster
Enschede, The Netherlands

Joshua Lieberman
Traverse Technologies, Inc.
Newton, Massachusetts

Roberto Lucchi
European Commission—Joint
    Research Centre
Ispra, Italy

Michael Lutz
European Commission—Joint
    Research Centre
Ispra, Italy

Paul van der Molen
ITC/Kadaster
Enschede, The Netherlands

Stefano Nativi
Italian National Research Council
Prato, Italy

Peter van Oosterom
Delft University of Technology
Delft, The Netherlands

Bo Overgaard
Grontmij-Carlbro
Glostrup, Denmark

Siva Ravada
Oracle Corporation
Nashua, New Hampshire

Marcel Reuvers
Geonovum
The Netherlands

Paul Scarponcini
Bentley Systems
Noida, India

Jayant Sharma
Oracle Corporation
Nashua, New Hampshire

Andrew Woolf
Rutherford Appleton Laboratory
Oxon, United Kingdom

Sisi Zlatanova
Delft University of Technology
Delft, The Netherlands

Introduction

*Paul Scarponcini, Styli Camateros, Oscar Custers,
Sisi Zlatanova, and Peter van Oosterom*

CONTENTS

Agreeing on high-level concepts of spatial data and the development of systems handling these is the first step towards spatial information infrastructures (SII). OGC and ISO/TC211 have developed a rich set of standards in this area (independent of specific themes or domains). Parallel to this development has been the growth of the Internet and all its protocols that have created the foundation of the SII. This does not mean that we understand each other's information, as for this we also have to agree on the domain (or thematic) models. In the context of these models the data get more meaning, and it is fair to state that data become information. Today these models are often expressed as Unified Modelling Language (UML) class diagrams, often limited to just the data side. The next step is applying knowledge and inference engineering technology. This chapter gives an overview of the different aspects of the SII. Next, attention is paid to the question of what is intended with the spatial semantic web. Finally, an overview of the spatial semantics evolution is described. Throughout this introduction references are made to the other chapters in this book, which provide further detail.

ASPECTS OF THE SII

Besides agreeing on the information content, there are a number of other aspects that are needed to realize the SII. These can be subdivided into technical and nontechnical aspects. Among the technical aspects, the SII needs metadata, described in catalogues and made available via registry services. The Web-based services (including data delivery services) form another technical aspect of the SII. Among the nontechnical aspects of the SII, the legal and organizational issues are important: copyright, pricing policy, access rights, etc. The BE 2007 research seminar focused on the technical aspects, and specifically those of the core element of the SII: the spatial information itself. Figure 1 gives an impression of how the information within the SII can be combined and used in various application contexts.

STANDARDIZATION OF THEMES: A NUMBER OF EXAMPLES

After the theme independent aspects of standardizing spatial information, there is now a wave of attempts to agree on complete themes, covering both the spatial and nonspatial aspects. A number of examples of standardizing themes/domains are now briefly presented. Within the (road) navigation sector, the Geographic Data Files (GDF, ISO/TR 14825:1996) have been developed by ISO/TC204 Intelligent Transportation Systems (ITS). More recently, "Recording and Exchange of Soil-Related Data" was submitted to ISO/TC190 Soil Quality. A third example is the submission by the International Federation of Surveyors (FIG) of the Core Cadastral Domain Model (CCDM) to ISO/TC211 (Lemmen and van Oosterom 2006). Note that the CCDM has recently been renamed the Land Administration Domain Model (LADM); see chapter 9. Also within the private sector, there are numerous development efforts for "domain" models, though the acceptance or integration of these into the de jure standards may be problematic.

Two important advantages of agreeing on domain models are (1) it becomes easier to understand the information of others within the domain and (2) system developments may be shared as many partners base their systems on the same model.

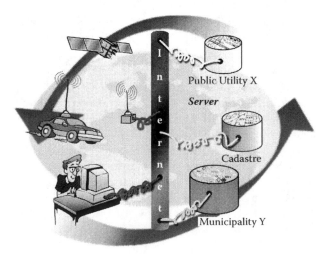

FIGURE 0.1   (See color insert.) Using the spatial information infrastructure.

The benefit of domain models (and ontologies) for facilitating information discovery and building knowledge-based systems is clear. The drawback of different ISO/TCs (or other organized sectors) for different geo-information themes is that there is no or difficult harmonization between themes (perhaps confusing overlap and also double work). Anyhow, it will not stimulate interoperability between these themes as needed for a wide spatial semantic web. The development of thematic (semantically meaningful) models is the future of geo-information standardization.

UNPRECEDENTED PROGRAMS: INSPIRE AND U.S. DHS GEOSPATIAL DATA MODEL

However, recently there have been a number of large initiatives that have started to develop harmonized (interoperable) model specifications covering many themes. For example, within INSPIRE, 34 different themes are to be covered; see chapter 1 and http://inspire.jrc.it/. It will be an incredible challenge for the 27 countries of the European Union to realize this: first agree on the harmonized models and next deliver information according to these models. But the situation is not unique for Europe; see the U.S. Department of Homeland Security (DHS) Geospatial Data Model, which also covers quite a broad number of themes, chapter 4, and http://www.fgdc.gov/fgdc-news/geo-data-model/. Parts of the DHS are based on the FGDC Framework Data Content Standard, which depends on the ISO TC211 feature model upper-level ontology, described in more detail in chapter 4. INSPIRE will also adopt various ISO TC211 standards. These large programs will create the infrastructure from which many applications and users will benefit, within government, the private sector, and individuals.

A MODEL-DRIVEN APPROACH

Creating the harmonized models and specifying them as UML class diagrams (and documenting them further with the help of feature and attribute catalogues) is in essence capturing (and agreeing on) human knowledge within a certain domain (or closed world). Note that UML is less suitable as a basis for creating bridges between domains (semantic mapping, translation, and transformations); other tools are needed for this, for example, OWL and reasoners. These UML class diagrams can be used for the implementation of information systems according to the model-driven architecture approach. The same model can be the basis of a database schema (SQL Data Definition Language), an exchange format (XML schema), or most of the user interfaces and associated behavior in an edit environment (e.g. automatically generating specific types of forms to enter valid attribute values of a specific feature). Within Bentley, the XFM technology is an important indication of this development.

THE SCOPE OF SII

Note that the SII not only covers traditional geo-information, but also (georeferenced) designs/models, and subsurface information (geotechnical, geological, etc.). Clearly, we have the issue of the 3D aspect in many of the relevant themes. Further, as things do change over time, the temporal element is also very important. How does this all fit into a usable interoperable infrastructure? The semantic aspect of information (what does it mean?) is not only important for human beings to understand each other, but semantics is also essential if we want machines to do useful things with that information. Therefore, the semantics will have to be formalized beyond what is currently expressed in UML, especially when trying to harmonize across domains. This is the essence of the semantic web: ontologies, perhaps expressed in OWL.

THE SPATIAL SEMANTIC WEB

So what exactly is the "Spatial Semantic Web" and how is this different from the proposed Semantic Web and even the current World Wide Web (WWW)? First, it is important to understand the limitations of the WWW.

THE WORLD WIDE WEB

Try to Google the first author's name, Scarponcini, in order to see all of the papers he has written. You will get close to 100,000 hits. Today's search engines merely look for keywords in documents, and, if found, return the entire document. It turns out that "Scarponcini" is Italian for a small (low-cut) boot. So most of the documents returned are in Italian and are about small boots. Searching instead for Paul Scarponcini will return a more manageable set of just under 1,000 documents, one of which is a story about *Paul* McCartney buying a pair of *small boots* in Italy. Yes, you could eliminate this one by putting quotes around the entire name. But what if a document contains only P. Scarponcini instead of Paul, or Scarponcini, Paul? The computer, the search engine, that is, has no idea that a successful find is a document that is a paper written by Paul Scarponcini. It also finds documents that are papers, but which are authored by someone else; Paul Scarponcini appears in the bibliography. Some documents are meeting minutes where Paul Scarponcini is in attendance. The other limitation of the current WWW is that it can only return (whole) documents. It is up to the user to read and interpret the result to see if there is anything of value in the document. The only information contained in the document (other than the document content itself) is the information the computer needs to be able to properly present the document on the screen. In its tremendous success in being able to make information readily available, the WWW may now be suffering because it returns just too much information for humans to consume.

THE SEMANTIC WEB

The next logical step then is to augment the information contained in documents with additional information that would allow the computer to understand the document content. This can include information about the document (metadata) as well as (ontological) information about the information contained in the document. A search engine could then make better decisions about what documents to return. Metadata and ontologies are discussed in greater detail in subsequent chapters, particularly metadata in chapter 10 and ontologies in chapter 3. The Web should also be able to provide more than just information in a document. Services will also be provided that can access, manipulate, integrate, and present information as the user wants to see it, instead of just how it appears in a single document. Searching for the right service poses an even greater challenge than searching for information, especially if multiple services need to be chained together to achieve the desired result (chapter 7). Additional information is available about the Semantic Web from various sources, so it does not have to be repeated in detail here. However, to summarize where we have been and where we may be headed, chapter 3 provides an assessment of significant achievements on semantics technologies to date, as well as important challenges that remain to be solved. What is germane to this book are the spatial aspects of the Semantic Web.

The Spatial Part

Almost every piece of information has a location associated with it. This might be the home address of a person, the business address of a restaurant, the current position of my car as I am driving down the road, or where the speed limit changes. One would therefore expect location information to play an important role in the Semantic Web. In fact, many have proposed using location as a means of integrating other information—see chapter 6. Augmenting the information that a restaurant only accepts cash with the locations of the restaurant and the nearest ATM might be helpful. But it is not quite that simple. There are numerous ways of expressing locations, from simple street addresses to more precise latitude and longitude coordinates. They can be absolute or relative, as two blocks down on the right (from your current location). So having the street address of the restaurant and the latitude and longitude of the ATM will not help unless you also have a way of correlating these two locations. What is special about spatial information that warrants its recognition in the Semantic Web? First, there are spatial types (point, line, polygon, polyhedron), but these can be described similar to nonspatial data using ontologies. Then there are fundamental spatial (intersects, within, touches) and topological (connected, adjacent) operators that require agreement on their precise meaning. These can then be augmented by more robust combinations using inferencing rules. For example, "near" might be defined using both distance and connectivity, such as determining if the ATM is near (within driving distance of) the restaurant. Spatial information is tied to a spatial referencing system that, in the case of geo-information, deals with the fact that distances between locations on the Earth's curved surface may not be the same as their distance apart on a flat map. Many indexing techniques have been developed for nonspatial or 1D information, but when 2D and 3D spatial representations need to be indexed, special methods are required. Additionally, much of the nonspatial data in the Semantic Web relates to discreet objects, which are easier to classify in an ontology. Many spatial objects, such as areal coverages and linear events, require special treatment both at the (continuous) object level as well as the location-dependent property value level (e.g. elevation map), adding to the ontological complexity. And for spatial data sets, relationships are more frequently computed rather than stored. This presents problems for reasoners, which assume that all relationships are explicit.

SPATIAL SEMANTICS EVOLUTION

Associating semantics with spatial is not a new concept. As early as the 1970s, ontologies were part of CAD, though not recognized as such. The Building Design System,

BDS, software from Applied Research of Cambridge, as well as the follow-up General Drafting System, GDS, required the user to identify an object before drawing any linework (McDonnell Douglas 1988). The object name was comprised of up to six facets, effectively creating a type hierarchy. When purchasing such a system, the user was faced with the daunting task of creating a relevant taxonomy of objects, such as BUILT:TRANS:ROAD:HIGHWAY:RT66:EAST for the eastbound carriageway of the Route 66 highway in the transportation domain of the built (vs. natural) environment. The benefit was the ability to then obtain a drawing of just those object classes desired, instead of the more rigorous layered approach common at that time. So you could get all the TRANS objects, no BLDGs, and highways could be rendered green whereas local roads could be brown and dashed. Properties could be attached to objects at either the object or instance level. Eventually, this nonspatial data could be stored in a relational database using SQL*CAD (Scarponcini 1989). This was taken to an extreme when GDS/SQL*CAD/Oracle were customized for the design of the $6 billion wastewater treatment plant in Boston (Scarponcini 1990). Object facets were used to identify both the facility being constructed as well as the system to be maintained and operated. All menus were data driven by queries to the database so that only the appropriate objects could be used for a particular facility. The resultant ontology for wastewater facilities was intended as a transition from design/construction into maintenance/operation. And this was in 1988. PCN data structures (for polygons, chains, and nodes) were added to GDS, and with the SQL*CAD connection to Oracle, GDS was a predecessor to feature-based GIS.

INTEGRATED DATA MANAGEMENT

Related to the data and operation model driven approach indicated above is the fact that the integrated management of data, both spatial and nonspatial, is the preferred approach. The GEO++ system (Vijlbrief and van Oosterom 1992) based on the Postgres DBMS (Stonebraker et al. 1990; de Hoop and van Oosterom 1992) is an early example of this approach. The spatial functionality within the DBMS is provided by Abstract Data Types (ADT) for point, line, and area geometry. Note that the spatial functionality was standardized before within SQL (ISO/IEC 1999; ISO 2002a). The interface of GEO++ is completely model driven based on the DBMS catalogues describing the defined tables and the available operators for each data type. This is true for both the 2D and the 3D version of GEO++; see figure 2. The Cadastral query tool is an example of a system based on the GEO++ model driven approach, though the research DBMS Postgres has been replaced by the production DBMS OpenIngres (van Oosterom et al. 2002).

AUTOMATED SPATIAL REASONING

To extend the semantics of GDS, the *Dafne* prototype was developed in 1991 (Scarponcini 1995). It captured spatial objects from GDS and their associated properties from Oracle, via SQL*CAD, and instantiated objects in Neuron Object, an artificial intelligence software package. Inferencing rules were added for reasoning about the spatial data. Sample CAD drawings and GIS maps were created in GDS and *Dafne* was able to infer information from the data that persisted. The near predicate

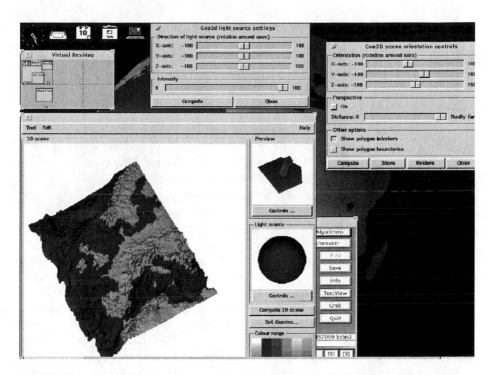

FIGURE 0.2   (See color insert.) Impression of the 3D version of GEO++, a model-driven GIS. (From van Oosterom, P. et al. 1994. *Proceedings of the International GIS workshop AGDM '94* [Advanced Geographic Data Modelling], Delft, The Netherlands, September 12–14, 80–95.)

mentioned earlier was implemented as a rule based on the geometry and topology of the roads on the GIS map or rooms on the CAD architectural plan. What made *Dafne* a challenge was that it predated any standards, like OGC and ISO TC211. There were no standard spatial data types, no fundamental spatial operators, and as yet no consensus on the validity of a feature-based approach. These were all proposed as part of the *Dafne* prototype, perhaps a bit ahead of its time.

SPATIAL STANDARDIZATION BEGINS

This all changed in the mid-1990s with the adoption of OGC Simple Features (OGC 1998), ISO/IEC SQL/MM Part 3:Spatial (ISO/IEC 1999), and ISO TC211 (ISO 2002a). The "simple" describes the geometries, not the features themselves, being limited to points, linestrings, and polygons with linear interpolation only and collections of these. Perhaps their greatest initial contribution was a (simple) geometry-type hierarchy (ontology), complete with properties and fundamental spatial operators. Perhaps most unknown at the time, the feature model that was specified in TC211 (ISO 2003) and supported by OGC and SQL/MM changed the fundamental basis of GIS. This is explained in chapter 4. A feature can have properties, including attributes, operations, constraints, and roles, as well as associations to other features. Of significance is the fact that spatial representation is considered to be an attribute of a feature, rather than being the central organizing type typical in early GIS—see chapter 2. The feature model allows for multiple geometric representations for the same abstracted real-world entity but offers no solution on how to "harmonize them." Chapter 8 elaborates on the need for semantic standardization.

Geometry- versus Object (Feature)-First Modelling Approaches

When we look at spatial modelling in the past, in principle we can distinguish between three different approaches: (1) geometry (topology) first, (2) object (feature) first, and (3) a hybrid approach. Because the related models have quite a different starting point, there is sometimes confusion between modellers.

In the geometry-first approach, the models start from the geometry (topology). Attributes are added to these geometries in order to classify the objects. The result is typically a set of tables in the database such as a point/symbol table, text/label table, line table, and area table. Within a table all objects (records) have the same set of attributes. For example, in the area table there may be houses and roads, all having the same attributes. In this approach, it is also possible to explicitly model the topological structure (e.g., linear network, or partition of space) with well-known advantages (explicitly connectivity, avoiding redundancy, better guarantees for quality under updates). The Dutch cadastral map in LKI is a typical example of this geometry (topology)-first approach (Lemmen et al. 1998). In this solution objects may share, via topology, their geometry with other objects. It could be argued that map representations (on paper or screen) themselves, that is, the visualization of the spatial data, are also a geometry-first type of model as all objects are considered together in a geometry model.

The second approach, the object-first approach, models the object classes first with added geometry. Every object class can have its own set of thematic attributes, which may vary for the different object classes. Also, every object has its own geometric description independent of any other object. The TOP10NL model is an example of this approach. Typically the result is a set of tables (or relational objects) in the database such as houses, roads, and waterways, which have among others their own simple object geometry type attribute. Sometimes additional rules (constraints) are added in order to avoid unwanted situations (often topology based); for example, a house polygon should not overlap with a road polygon at the same level/layer. The drawback is that all these constraints have to be explicitly stated (and checked when updates are performed) and are not embedded in the main structure of the object-first type of model. Also, the model does not explicitly contain the topological relationships, which may support various types of analyses (e.g., quality control of updates). It must be noted that topological relationships are very important for map generalizations; for example, what are the neighbors of this object (candidates for aggregation), is the network connectivity damaged when this road segment is removed, etc. The ISO 19109 standard makes properties (including geometries and topologies) composites of features, thereby precluding their sharability.

The third approach is the hybrid approach, which treats the geometry and object class equally. It combines the strengths of both approaches: the (thematic) attributes are specifically designed for every object class, but the model also enables shared geometry and use of embedded structure. The spatial domain is partitioned and the result is described using tables for nodes, edges, and faces (and solids in three dimensions). The objects are modelled in the same way as in the object-first approach, with the exception that the objects do not have their own independent geometry-attributes, but refer to primitives in the geometry/topology part of the model (node, edge, face). This is the approach described in the formal data structure (FDS) theory of Molenaar (1989) and quite recently implemented in products such as 1Spatial (LaserScan) Radius Topology and Oracle Spatial Topology (first introduced in version 10g). It is also supported by the new topology part of the SQL/MM standard and implemented in the ISO TC204 GDF standard. Unfortunately, the security system enforced by the SQL language (i.e., GRANTs) did not allow the topological primitives to be implemented as relational objects like the geometry types, so tables were used instead in order to enable sharability of topology. The OGC Simple Feature standard only supports geometry, and sharing it between features is not precluded, as is the case with SQL/MM geometry.

De Hoop et al. (1993) discuss the different modelling approaches and the consequences for realization and use. It cannot be claimed that one model is better than another model. This depends on the application context and use. If one specifies a number of important characteristics of the application domain and typical use, then it is possible to state which approach is preferred. Considerations could be

1. Allow exceptional overlapping of objects in certain cases (e.g., bridge over water in two dimensions)
2. Allow modelling of systematically overlapping sets of object classes (e.g., topographic objects on one hand and administrative units on the other hand)
3. Enable multiple geometry representations of single objects (e.g., road area polygon and road center line, or building footprint polygon, building rooftop polygon, and building centroid)
4. Support consistent updating/maintenance
5. Support efficient querying, analysis, and viewing of data
6. Avoid storage space consuming representations (might also be expensive for data transfer)
7. Support easy delivery for customers (simple objects might be easier to receive in a system other than topology structure), etc.

STANDARDS PROGRESS

After Simple Features, OGC was successful in launching the WMS, Web Map Service (OGC 2004). Chapter 11 presents the comprehensive Danish strategy for SII based on WMS and augmented by various registries. WMS, though service-based and web enabled, suffers from the same limitations mentioned earlier for the WWW—the map, like a document, is retrieved in its entirety, with no knowledge of its content. So now the standards seem to be in a bit of a free fall. OGC has proposed WFS, Web Feature Service (OGC 2005), a more semantically oriented version of WMS, and a WCS, Web Coverage Service (OGC 2006), and the resultant confusion over whether a coverage is a map or a feature. This has led to the creation of an Architecture Board to help sort out a conceptual model for a family of Web service specifications but fundamentally a strategy for writing specifications and, even more fundamentally, identifying the mission of OGC. TC211 appears to be equally astray. The initial 20 domain-neutral standards in the ISO 191nn family presented a concise, consistent abstract model of all aspects of geographic information. The value of their contribution of the feature model as an upper-level ontology suitable for the spatial semantic web cannot be underestimated. But the subsequent 30 standards appear to be all over the place, including domain-specific topics such as transportation tracking and navigation. Perhaps it is appropriate that standard number 19150 is focused on semantics (ISO 2007). Some of the challenges beyond simple features and the feature model are now being wrestled with, both in OGC and in TC211. This includes observations and measurement, temporal, and services. Chapter 5 focuses on earth sciences. Here, it is often inadequate to model the data resulting from an observation without details of the instrument or process used to generate the data. OGC is working on reconciling an abstract model of observation and measurements with their current GML implementation. TC211 has managed a temporal schema, ISO 19108 (ISO 2002b), defining fundamental temporal types and operators. Their Moving Features standard, ISO 19141 (ISO 2005), considers changes in location over time. The more difficult problem of features morphing (changing their physical characteristics) over time has yet to be standardized. Chapter 7 transitions us into the world of services, identified by the operations they support. Each operation can be defined by its input and output parameters, which are conveniently based on the aforementioned feature model. The whole semantics of individual services and service chaining is certainly in the research phase, though solutions to nonspatial service semantics may provide hints to solve spatially enhanced services.

REFERENCES

de Hoop, S., L. van de Meij and M. Molenaar, 1993, Topological relations in 3D vector maps. In: *Proceeding of 4th EGIS,* 448–455. Genoa, Italy.

de Hoop, S. L. and van Oosterom, P. 1992. Storage and Manipulation of Topology in Postgres. In *Proceedings EGIS'92*, Third European Conference on Geographical Information Systems, 1324–1336. Munich, Germany.

ISO 2002a. *ISO IS 19101:2002, Geographic Information—Reference Model.* Geneva: International Organization for Standardization.

ISO 2002b. *ISO IS 19108:2002, Geographic Information—Temporal Schema.* Geneva: International Organization for Standardization.

ISO 2003. *ISO IS 19109:2003, Geographic Information—Rules for Application Schema.* Geneva: International Organization for Standardization.

ISO 2005. *ISO CD 19141, Geographic Information—Schema for Moving Features.* Geneva: International Organization for Standardization.

ISO 2007. *ISO New Work Item Proposal, Geographic Information—Ontology.* ISO document 211n2163. Geneva: International Organization for Standardization.

ISO/IEC 1999. *ISO/IEC IS 13249-3:1999(E) Information Technology: Database Languages—SQL Multimedia and Application Packages—Part 3: Spatial.* Geneva: International Organization for Standardization.

Lemmen, C. H. J., E. P. Oosterbroek, and P. J. M. van Oosterom. 1998. Spatial Data Management in the Netherlands Cadastre. In *Proceedings of the FIG XXI International Congress,* Commission 3, Land Information Systems, 398–409, Brighton, United Kingdom, July 1998.

Lemmen, C. and P. van Oosterom. 2006. Version 1.0 of the FIG Core Cadastral Domain Model, *XXIII International FIG Congress*, Munich, October 2006.

McDonnell Douglas. 1988. *GDS Reference Manual*, Release 4.10. McDonnell Douglas Corporation, St. Louis, MO.

Molenaar, M. 1989. Single Valued Vector Maps: A Concept in Geographic Information Systems. *International Journal of GIS* 2 (1):18–27.

OGC. 1998. *OpenGIS® Simple Features Specification for SQL Revision 1.0.* Wayland, MA: Open GIS Consortium, Inc.

OGC. 2004. *OGC Web Map Service Interface,* Version 1.3, OGC 03-109r1. Wayland, MA: Open GIS Consortium, Inc.

OGC. 2005. *OGC Web Feature Service Implementation Specification,* Version 1.1, OGC 04-094. Wayland, MA: Open GIS Consortium, Inc.

OGC. 2006. *OGC Web Coverage Service (WCS) Implementation Specification,* Version 1.1, OGC 06-083r8. Wayland, MA: Open GIS Consortium, Inc.

van Oosterom, P., W. Vertegaal, M. van Hekken, and T. Vijlbrief. 1994. Integrated 3D Modelling within a GIS. In *Proceedings of the International GIS Workshop AGDM '94* (Advanced Geographic Data Modelling), Delft, The Netherlands, 12–14 September, 80–95.

van Oosterom, P. J. M., B. Maessen, and C. W. Quak. 2002. Generic Query Tool for Spatiotemporal Data. *International Journal of Geographical Information Science* 16 (8):713–748.

Scarponcini, P. 1989. CADD/DB: Fourth Generation Application Platform for Software Integration. In *Computing in Civil Engineering: Computers in Engineering Practice, Proceedings of the Sixth Conference*, American Society of Civil Engineers, New York.

Scarponcini, P. 1990. CADD/DB: The Boston Harbor Cleanup Project. *1990 International Oracle User Week Proceedings*, Anaheim, CA, September.

Scarponcini, P. 1995. An Inferencing Language for Automated Spatial Reasoning about Graphic Entities. In *Advances in Spatial Databases,* Lecture Notes in Computer Science no. 951, ed. M. J. Egenhofer and J. R. Herring, 259–278. New York: Springer. Fourth International Symposium on Large Spatial Databases—SSD'95.

Stonebraker, M., L. A. Rowe, and M. Hirohama. 1990. The Implementation of Postgres. *IEEE Transactions on Knowledge and Data Engineering* 2 (1):125–142.

Vijlbrief, T. and P. van Oosterom. 1992. GEO++: An Extensible GIS. In *Proceedings 5th International Symposium on Spatial Data Handling*, Charleston, South Carolina, 40–50.

1 Requirements and Challenges for Building a European Spatial Information Infrastructure: INSPIRE

Alessandro Annoni, Anders Friis-Christensen, Roberto Lucchi, and Michael Lutz

CONTENTS

The Sixth European Environment Action Programme emphasizes the need to base environmental policy on sound knowledge and participation, principles that will influence the European Union's environmental policy decisions for the next decades. In response to this and other high-level European Union strategies, the Infrastructure for Spatial Information in Europe (INSPIRE) initiative was conceived in 2001. Following

3 years of intensive collaboration with member states' experts and stakeholder consultations, on July 23, 2004 the Commission adopted a proposal for a directive for establishing an infrastructure for spatial information in Europe (INSPIRE). After the formal approval by the Council and by the European Parliament, the INSPIRE Directive 2007/2/EC came into force on May 15, 2007. INSPIRE will be built on top of national spatial information infrastructures* established and operated by member states with the purpose of supporting environmental policies and policies or activities that may have a direct or indirect impact on the environment.

This chapter describes many of the challenges encountered in building INSPIRE. Section 1.1 describes the background to, the key objectives of, and the current process being established for the development of the technical components of INSPIRE. Also, the link between INSPIRE and international standardization initiates related to spatial information infrastructures (SIIs) is discussed briefly. Section 1.2 presents some general functional use cases that an SII needs to support. Section 1.3 describes the challenges in building INSPIRE, many of which can be considered to be "semantic." Finally, section 1.4 summarizes the chapter.

1.1   BACKGROUND

There is a growing awareness that we are living at a time when environmental changes have a known and increasing impact on our economy and social well-being, and some require urgent action. Understanding the complex interactions between natural and human systems requires easier access to reliable and timely spatial information. In order to do so we should overcome key barriers still affecting Europe, including

*   Inconsistencies in spatial data collection—spatial data are often missing or incomplete, or the same data are collected twice by different organizations
*   Lacking documentation—description of available spatial data is often incomplete
*   Spatial data sets not compatible—spatial data sets often cannot be combined with other spatial data sets
*   Incompatible geographic information initiatives—the infrastructures to find, access and use spatial data often function in isolation only
*   Barriers to data sharing—cultural, institutional, financial and legal barriers prevent or delay the sharing of existing spatial data

These were the reasons given in the Memorandum of Understanding between Commissioners Wallström, Solbes, and Busquin signed on April 11, 2002 for developing the INSPIRE initiative.

From the outset of this initiative it was recognized that to overcome the barriers highlighted above it would be necessary to develop a legislative framework requiring member states to coordinate their activities and agree on a minimum set of common standards and processes.

Since May 2007, INSPIRE has been the legal framework (Commission of the European Communities 2007) to be implemented throughout the European Union

---

* We consider spatial information infrastructure to be synonymous with spatial data infrastructure.

with different types of geographical information gradually harmonized and integrated, resulting in a European Spatial Information Infrastructure.

1.1.1 REQUIREMENTS

INSPIRE lays down general rules to establish an infrastructure that will build upon infrastructures for spatial information established and operated by the member states. Implementing rules are needed for each of the key components of the infrastructure, namely: metadata, data specifications and harmonization, network services, data and service sharing, and monitoring and reporting. Given the political context of the directive, their drafting requires not only a high level of technical competence, but also the full participation and engagement of all the key stakeholders in geographic information in Europe. The general background and the INSPIRE requirements are described in Bernard et al. (2005).

To organize this process two mechanisms have been put in place: the first is to engage the organizations at the European national and subnational levels that already have a formal legal mandate for the coordination, production or use of geographic and environmental information. The second is to facilitate the self-organization of stakeholders, including both data providers and users of spatial data, into Spatial Data Interest Communities by region, societal sector and thematic issue. Involving all the interested parties from the very beginning and giving them leading roles in shaping the infrastructure is considered a key success of the INSPIRE process, as described in Craglia et al. (2003) and Masser (2007). The INSPIRE directive (Commission of the European Communities 2007) came into force on May 15, 2007 and is due to be adopted in the member states not later than May 2009. The directive foresees its implementation through the adoption of implementing rules following a specific timetable. For example, the metadata for the annex I and II spatial data themes (e.g., coordinate reference systems, geographical names, and land cover) shall be created by the member states in 2010 should the metadata implementing rules be adopted in 2008.

At the time of this writing, draft implementing rules for the metadata, network services, and monitoring chapters have been released by the relevant drafting teams for review and further processing. The data drafting team has released the draft of several documents defining a framework for the future Spatial Data Sets and Services Implementing Rules: the definition of the Annex Themes and Scope, the Generic Conceptual Model, and the Methodology for the Development of Data Specifications.

1.1.2 INTERNATIONAL ISSUES

The implementation of INSPIRE should be supported by international standards and standards adopted by European standardization bodies. In this way INSPIRE will benefit from the state of the art and actual experience of information infrastructures. Accordingly, the INSPIRE implementing rules should be based, where possible, on such standards and should not result in excessive costs for member states. As a consequence, the development of INSPIRE is influenced by and influences the development of de jure (ISO TC 211, CEN TC 287) and de facto standards (e.g., OGC).

In addition, INSPIRE is seen as the European contribution to other global initiatives such as GSDI (Nebert 2004), UNSDI (UNGIWG 2007) and GEOSS (a worldwide effort to build a Global Earth Observation System of Systems) (Battrick 2005). Both GEOSS and INSPIRE are distributed systems of systems built on international cooperation among existing observing and data management systems enabling the collection and distribution of accurate, reliable spatial data; information; products; and services in an end-to-end process.

1.2   USE CASES

INSPIRE does not require the collection of new spatial data, so emphasis is given to interoperability and harmonization of existing heterogeneous available information.

Based on the INSPIRE requirements, we identify the following general functional use cases that a European Spatial Information Infrastructure needs to support:

- Discover resources—data and services: It must be possible to discover resources—data as well as services—within the community based on the metadata provided. These metadata need to describe the content of and how to access data and services.
- Access data: In order to view the discovered data or use them in further analyses, it has to be possible to access the data. Data access has to be provided for both vector and coverage data.
- Use data: The interpretation of the content of data is necessary in order to use the data. To facilitate the utilization of data, metadata should be available to assess its suitability for the purpose. These metadata should describe all aspects of the data, for example, access constraints, quality, lineage, and intended usage.
- Visualize data: For illustration purposes, data need to be visualized and portrayed according to given rules (symbology). Objects can be viewed at different scales; for example, a building can be visualized as a polygon at one scale and as a point at another.
- Harmonize and integrate data: In particular, for cross-border data access there is a need to harmonize and integrate data coming from different sources with different application schemas. This requires the ability to transform data from a source schema to a given target schema.
- Orchestration: Since the functional use cases, for example, access, use, harmonization and visualization, may require the interaction of several services, there is a need to control the invocation of services.

1.3   TECHNICAL CHALLENGES

In the following section, we describe some of the challenges that we consider pertinent in order to build a European Spatial Information Infrastructure having the use cases described above as a basis. The challenges are interrelated, and some may be the cause of others. We describe each of these challenges in detail and point out related work in the area and directions for possible solutions. In particular, we go into detail with *inconsistent data* and *multilinguality*, as these are challenges that are of specific importance when creating a Spatial Information Infrastructure at the European level.

1.3.1 INCONSISTENT DATA

Data spanning across regions and national borders have inherently independent models for capture, storage and interpretation, as each region or national member state may have its own requirements and data are collected based on different and independent application needs, as well as different cultural heritages (e.g., Napoleonic cadastre). When accessing data across borders, this inconsistency becomes evident, for example, that roads meeting across a national border do not connect as the creation of roads is based on different practices. Another example is soil maps that show different soil types (from the *same* classification system) on both sides of a state border. This thematic inconsistency is caused by the fact that the data in both states were produced by different surveyors who subjectively decided on the soil type in a given area. An example of a geometric inconsistency is depicted in figure 1.1, where the geometries of roads do not meet at cross-border areas.

As data inconsistency is a broad topic and is caused by a number of different aspects, we list some important causes, especially when operating with geographic data. Several of them we revisit when describing further challenges, such as multilinguality, classification systems, reference systems, units of measure, and quality of data. The list is not complete and is based on descriptions in Kashyap and Sheth (1996) and Stuckenschmidt (2003):

Naming conflicts (semantics) occur when classes or attribute types with different semantics are given the same names (homonyms) or when classes or attribute types that are semantically the same are named differently (synonyms). The latter occurs in nearly all cases between the member states, because most use their native language.

Scale conflicts occur when attribute values have different units of measure or are represented in varying scales of measure, for example, nominal, ordinal or ratio.

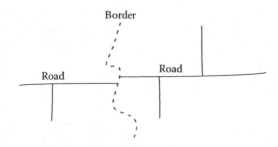

FIGURE 1.1 Example of a cross-border scenario where the geometries of a road do not meet at the border.

Precision or resolution conflicts occur when attribute values have different resolutions and precisions, for example, if two similar measurements are made with sensors with different precisions.

Constraint conflicts occur, for example, when two databases have different integrity or data capture constraints. An example is that roads under a certain size are registered in one database but not in another.

Data value conflicts occur when attribute values of objects that are represented in cross-border areas have different values. An example is geometry, as depicted in figure 1.1, but it could also be a thematic attribute such as the number of road lanes.

Inconsistent data complicates the integration and harmonization of two or more data sets as additional processing is required. For example, in order to connect two roads (as seen in figure 1.1) some geometric operations are necessary. Harmonization is further complicated if the data to be integrated are inconsistent in themselves (i.e. if constraints are violated). An example could be a road object type that has an attribute road class indicating the type of road restricted by a domain: main, local and private. If an attribute value is set to "highway," this violates the constraint and, thus, produces an inconsistent data set.

To facilitate integration and harmonization, transformations among inconsistent data and schemas are required. Some of the challenges have been addressed, for example, in the GiMoDig* and SDIGER† projects. The objective of the GiMoDig project was to develop methods for delivering geospatial data to a mobile user by means of real-time data integration and generalization (see, e.g., Afflerbach, Illert and Sarjakoski 2004) and its proposed methodology was reused in the SDIGER project. However, GiMoDig only addressed the integration of small subsets of data used for handheld devices and the proposed methods and technologies need to be revisited in order to validate its feasibility for larger scale integration such as at the pan-European scale. Another challenging research area is how to do (semi)-automated generation of schema mapping rules to perform a schema transformation. Automation of schema mapping has been under investigation for more than a decade and poses a number of challenges; for example, Patil, Dutta, and Sriram (2005) and Sheth, Gala and Navathe (1989) both state that schema mapping cannot be completely automated. For a comprehensive survey and perspectives of schema matching and integration, see Patil, Dutta, and Sriram (2005). The idea is that instead of manually having to define mappings and functions in transforming data in one schema into another, they should be generated automatically or at least some guidance should be provided in the generation. Several approaches can be taken, for example, simple linguistic mapping, where names of attribute types are matched (Madhavan, Bernstein, and Rahm 2001). However, such an approach does not solve the above-mentioned conflict types and more advanced approaches, for example,

---

* GiMoDig: Geospatial info-mobility service by real-time data integration and generalization. http://gimodig.fgi.fi.

† SDIGER: The SDIGER project consists in the development of a Spatial Data Infrastructure (SDI) to support access to geographic information resources concerned with the Water Framework Directive (WFD). http://www.idee.es/sdiger.

based on ontologies, could be used in order to match to heterogeneous schemas. A survey and comparison of existing approaches to schema matching, ranging from simple linguistic approaches to more advanced approaches, is found in Shvaiko and Euzenat (2005).

1.3.2 MULTILINGUALITY

Multilingual aspects relate to almost all functionality envisaged within a European SII. They concern the entire range from the translation of standards to the interpretation of the schemas and the content of geospatial data and metadata. This can complicate the discovery and access to data, for example, if a requester uses search terms (e.g. keywords) in a language different from the one used in the metadata created by the data provider. Language-specific geographic names (or toponyms), such as Roma (official name) versus Rome (exonym), that can be used for characterizing the spatial extent of an object in a catalogue query can be considered a special case of this problem. Similar problems can occur during data access when terms used in a query expression use a different language from the one used in the data. Multilinguality can also complicate the harmonization and integration of data. Depending on the specific requirements for the harmonization (e.g., pan-European reporting), the content might have to be translated to a specific target language. Finally, visualization of data can be affected, even though to a large extent it can be language-independent. Nevertheless, for the display of legends, attribute information, and the labelling of features, multilingual issues should be considered.

In the INSPIRE context where data from multiple member states have to be handled, the solution to multilingual issues is not to translate everything into a common language (e.g., English). Rather, depending on the context and application, the following strategies and approaches can be adopted to overcome these issues.

To facilitate discovery in a multilingual environment, the user interfaces of clients and applications, including portals intended for an international audience, should be internationalized (see examples in Ostländer, Tegtmeyer, and Förster 2005; Tchistiakov et al. 2005; Zarazaga-Soria et al. 2006) and strategies for cross-language information retrieval (of data and metadata) should be developed (Nogueras-Iso et al. 2005). The latter could include the automatic translation of queries to all supported languages, the automatic translation of metadata documents to all supported languages, or indexing document and queries in some common and language-independent representation. In general, automatic translation tools, such as the rule-based *Systran* (Senellart, Dienes, and Váradi 2001) or the statistics-based *LanguageWeaver* (Knight and Marcu 2005; Benjamin, Knight, and Marcu 2002) software do not provide 100% accuracy (meaning that both translations can be unambiguously interpreted). However, at least they can help the user to better understand the metadata descriptions or user/technical specifications that were written in a different language to that required by the user.

To support translations, controlled lists, such as those defined in ISO 19115 (ISO/TC-211 2003c) and (multilingual) thesauri should be used instead of free text attributes in application schemas (and therefore also in queries). In order to help users and applications to understand what can be expected in metadata records created in other countries with different natural languages, a description of these controlled lists and a central thesaurus for the translation would be required.

Existing multilingual taxonomies, thesauri and ontologies include GEMET (GEneral Multilingual Environmental Thesaurus, http://www.eionet.eu.int/GEMET), EDEN-IW (Environmental Data Exchange Network for Inland Water, http://www.eden-iw.org/) (Felluga and Plini 2004; Lucke et al. 2003) and AGROVOC (http://www.fao.org/agrovoc/) (Soergel et al. 2004). In an SII, such thesauri should be made available as Web services, such that they can be queried by other services and/or clients. The SKOS (Simple Knowledge Organisation System) Core specifications (Miles and Brickley 2005) can be used for expressing the basic structure and content of concept schemas (thesauri, classification schemes, subject heading lists, etc.).

For overcoming problems caused by different geographic names used in different languages, multilingual gazetteers can be used. In the EuroGeoNames (EGN) project, a Web service will be implemented that provides access to the official, multilingual geographical names data held at the national level across Europe (Sievers and Zaccheddu 2005). A more bottom-up approach is taken by the geonames.org (http://www.geonames.org/) database and Web services, which are available under a creative commons attribution license. It is based on existing lists of geographic names, for example, by the U.S. National Geospatial-Intelligence Agency (NGA) or wikipedia.org, and can be manually edited, corrected and extended by users through a wiki interface.

Problems during the harmonization and integration of spatial data are often caused by different conceptualizations between different language or information communities. They cannot be solved using simple translations and term mappings, but might require more heavyweight semantic approaches (see also the previous section on inconsistent data).

1.3.3 Multiple Representations

A geographic object can be defined differently depending on the universe of discourse and in the resulting conceptual schema; in other words, the representation of an object is application-dependent. An example can be roads, which in a topographic map are defined as surfaces, while in a route planning system, they are defined as centerlines constituting a network. This complicates harmonization because this requires well-defined models to describe the relationships among the multiple representations of a single real-world entity. Furthermore, it complicates access and visualization because a user will have to specify which given representation and visualization he wants to access. Multiple representations can also be intended to provide solutions in situations where objects are needed at different scales (hence the term *multiscale database*). In this situation, there typically is a reference (or base) scale from which all other objects are derived (Kilpelainen 1997). In certain cases, objects are derived once from a reference scale and then any linkage is deleted. This means, however, that whenever objects at the reference scale are updated, a complete derivation of objects is necessary. It is more efficient to keep a linkage among objects (the different representations) in order to avoid having to recreate all objects whenever there is a change at the reference scale. This requires a multirepresentation database where rules that specify interdependencies can be stored to avoid inconsistent representations when one object is updated (Devogele, Trevisan, and Raynal 1996; Rodríguez and Egenhofer 2003; Friis-Christensen et al. 2005). Such systems already exist and are being used; for example, LaserScan LAMPS2 and ESRI ArcGIS have features for handling multiple representations. An alternative to multiscale and multirepresentation databases is to address multiple representations by a variable-scale approach as described in, for example, van Oosterom (2005). Still, most approaches are tool-dependent, and well-established models describing relationships among representations would ease harmonization and integration.

1.3.4 CLASSIFICATION SYSTEMS

Data content is often based on classification systems (e.g. for land use or soil types), which depend on application-specific requirements or national standards, an example being the Corine Land Cover classification system (Commission of the European Communities 1995). Classification systems can be used as controlled vocabularies for data or metadata attributes and thus facilitate discovery or access. However, they can also cause problems in these tasks if the requester uses terms from a classification system different from the one used in the (meta)data. If the classification system used is not documented, this can also complicate data interpretation. Finally, when trying to harmonize or integrate data using different classification systems, a mapping between these (or to a common target system) has to be found. For example, a data set with a classification system containing *coniferous forest* and *deciduous forest* classes can be mapped to a data set with a classification system containing a single class *forest*.

Many approaches have been proposed to map between different classification systems. Broadly, these approaches can be divided into two groups. *Ontology-based approaches* (e.g. Lutz et al. 2007; Visser and Stuckenschmidt 2002; Visser et al. 2002; Visser, Vögele and Schlieder 2002; Vögele and Spittel 2004) build on formal definitions of concepts in ontologies, for example, using Description Logics (DL) (Baader and Nutt 2003), and reasoning on these concepts. These concepts define the classes in the classification systems to be mapped. To be able to compare the concepts, the definitions should be based on shared vocabularies (Wache et al. 2001) containing the basic characteristics underlying the different classification systems. For example, in the GEON project (Chen et al. 1996), geological age, composition, fabrics, texture and genesis were used as a simple shared vocabulary. These terms were used to define concepts describing the terms used in the geological classification systems of different U.S. states, and to integrate (based on DL reasoning) data using these different classification systems in a single map. Also, *similarity-based approaches* (see Goldstone and Son 2005 for an overview and Raubal 2004; Gratius, Bertran and Rodriguez 2004) for example applications in the geospatial domain) are based on basic terms to describe concepts. However, rather than deriving a taxonomy between concepts, they compute a numeric similarity value between two concepts that can express gradual differences between them. In the HarmonISA project (Hall 2006), a similarity-based approach is used to combine land use data from the border region of Austria, Slovenia and Italy in a single land use map. In this project, a comprehensive shared vocabulary for defining land use classes has been developed based on the different national as well as the European CORINE land use classification systems. The heterogeneity that can exist between different classification systems is illustrated in figure 1.2, which shows parts of the structure of the Austrian Realraumanalyse and the European Corine land use classification systems.

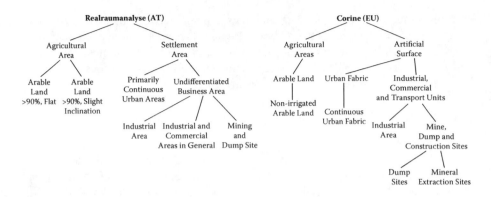

FIGURE 1.2   Excerpt from the structure of the Austrian Realraumanalyse (left) and the European CORINE land use classification systems. (Based on Hall 2006.)

1.3.5   REFERENCE SYSTEMS AND UNITS OF MEASURE

Data containing spatial locations, time stamps or measurement values have to refer to a reference system or unit of measure in order to allow unambiguous interpretation. Which reference systems or units of measure are used can differ considerably by country, application or domain. Particularly the choice for a projected coordinate reference system depends on the data's absolute position on the globe and on the data's intended use (e.g. navigation, area or distance calculation). Figure 1.3 shows the differences in the representation of areas in different projections.

Data harmonization and integration require a full description of the reference systems or unit of measures used for all data to be integrated. In case they do not match, a transformation into a common reference system or unit of measure is required. For visualization purposes, this is particularly crucial, as all data sets to be visualized on one map have to share a common coordinate reference system.

(a) Mollweide Projection   (b) Mercator Projection

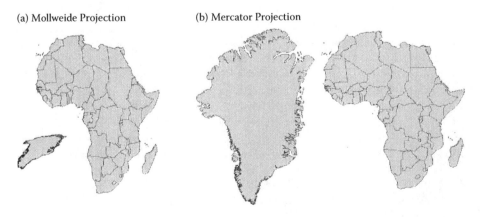

FIGURE 1.3   Greenland and Africa shown in (a) the equal-area Mollweide projection (showing the true area proportions) and (b) the Mercator projection (which strongly overrepresents areas towards the poles). (Example taken from Furuti 2003).

To handle these issues in a European SII, the units of measure and reference systems used in a specific data set should be unambiguously documented in its metadata. Units of measures and reference systems, as well as rules for their transformation, have to be described and stored in a central registry. This description should also cover error information in the event a transformation between two reference systems or units of measure causes a loss of precision. An example is transformations between coordinate reference systems that require a change of the reference datum. These transformations are always approximations and therefore should be known to the user in order to judge the loss in precision. But coordinate conversion between coordinate reference systems that are based on the same datum can also produce inappropriate results, if the target coordinate reference system is not suitable for the area in which the data to be converted are located (e.g., if an inappropriate Universal Transverse Mercator [UTM] zone is chosen). SII should also offer services to execute these transformations. How these are to be combined with existing service types for data access (in a loosely or tightly coupled manner) is currently an open research issue (Friis-Christensen et al. 2007).

1.3.6 OBJECT IDENTITY AND OBJECT LIFE CYCLE RULES

A unique object identification is necessary for data exchange purposes. When different institutions and authorities need to exchange data, unique object IDs are useful. Furthermore, object identity is needed in order to provide a framework for references of information, which is highly useful in the public administration when combining, for example, different registries and map databases. A possible solution to this is to establish an authoritative object ID supplier that provides object identifiers upon request following a standard scheme. Object identifiers are well known in the information technology domain under names such as Universal Unique Identifier (UUID) and Global Unique Identifier (GUID). There are also well-known schemes for representing UUIDs; see, for example, Leach (2005). Within the geographic information domain, the necessity of unique object identifiers has been described in, for example, Bishr (1999) and Sargent (1999).

In order to have a consistent object ID model, rules are required for determining the evolution of objects and, thus, their IDs. Consider the example that a building object is extended by a garage. In the database representation of the building, there are several possibilities: (1) the object is updated (extended) and the object ID maintained; (2) one new object is created and assigned a new ID; and (3) two new objects are created, one having the old ID, one assigned a new ID. This example is depicted in figure 1.4.

Different schemes for object IDs and life cycle rules complicate interpretation and harmonization among multiple data sets, as different identification mechanisms have to be used. Life cycle rules have been described, for example, by the Ordnance Survey as part of their distribution of the OS MasterMap (Ordnance Survey 2006). One of the specific challenges is how to model the life cycle rules at the conceptual and logical levels, and even more challenging is how to incorporate (or connect) specification of life cycle rules in the modelling of features, for example, in application schemas compliant to ISO 19109 (ISO/TC-211 2003a).

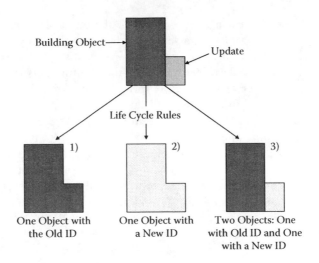

Building Object⟶

Update

Life Cycle Rules

1)   2)   3)

One Object with   One Object with   Two Objects: One
the Old ID   a New ID   with Old ID and One
                                          with a New ID

FIGURE 1.4   Various object life cycle rules applied during an update of a building object.

1.3.7   DATA QUALITY

The quality describes how "good" the data are in a certain context. Such information is important when assessing the credibility of data for usage and should be considered mandatory in metadata. The notion of data quality has some fundamental objective components, such as accuracy and precision. These components can be measured against application needs in order to define application-dependent meanings of data quality, which is also called the "fitness for use," thus allowing for a given application to answer the question: what is the quality of the data? Data quality and geographic data quality have been subject to research for several decades (Wang, Ziad, and Lee 2001; Guptill and Morrison 1995; Goodchild and Gopal 1989; Veregin 1999) and the various elements of data quality have been thoroughly described. Also, the International Organisation for Standardisation (ISO) has published standards addressing geographic data quality (ISO/TC-211 2002, 2003b).

Data quality complicates the harmonization among multiple data sets as varying levels of quality (or different quality definitions) may hinder integration. Furthermore, it may complicate the interpretation of several data sets if quality or the schema of quality descriptions varies such that comparisons are difficult or impossible. This calls for harmonized description models of quality that enable comparisons between different data sets. The ISO standards for quality (ISO/TC-211 2002, 2003b) can help in this process. However, because the standards are very broadly specified (in order to encompass all kinds of data), the definitions of some elements are not very clear or overlapping. This can lead to different interpretations of the quality elements, for example, how can *conceptual consistency* (adherence to rules of the conceptual schema) be distinguished from *completeness* (objects are present or overrepresented)? An approach to streamline description models of quality could be profiling of quality schemas for certain types of data. Additionally, the methodologies that are needed in order to assess quality should also be clearly defined and harmonized.

As distributed geoprocessing will be a likely future activity supported by SIIs, an additional challenge is how errors will propagate when being processed and combined with multiple sources. Some work in error propagation in spatial modelling and distributed models may be used as a basis for developing future approaches in this research area (Heuvelink 1998; Phillips and Marks 1996).

1.3.8 FUNCTIONALITIES (SERVICE OPERATIONS)

The service interface describes how to use a service, but does not cover semantic aspects of exhibited operations (i.e. what functionality is supplied by the service). In order to also describe this aspect of a service, its metadata should include information on the functionality. Currently, service functionalities is a very debated topic and the most credited solutions are based on ontologies; see, for instance, OWL-S (Smith et al. 2004), WSMO (Fensel et al. 2007) and SA-WSDL (Farrell and Lausen 2007).

Service functionalities play a central role during the discovery where they can be used to verify whether a service supports the required functionality. As far as orchestration is concerned, the combination of service functionalities raises harmonization and integration problems similar to the ones that occurr when data are combined. This means that, in order to enable service composition, the orchestration should foresee a method for transforming services able to deal with aspects like, for instance, different versions of standards and different input and output schemas. Solutions have been proposed that are based on introducing additional services (called mediators or adaptors) in the workflow whose role is to make the services chained in the orchestration interoperable (see, e.g., Brogi, Canal, and Pimentel 2006; Gagnon 2007; Lemmens 2006).

Another aspect of orchestration is the quality of the results, which could be affected by the way that the overall functionality is achieved. Consider the case of an orchestrating service that translates words from and to different languages. In order to accomplish this, it uses two services that are able to translate from any languages to English and from English to any other languages, respectively. This double translation that occurs when we pass from one language via intermediate results in English to a target language (e.g., translating from French to German) is likely to reduce the quality of the results compared to a direct translation between the two languages. Finally, it would be interesting to investigate whether such functionality descriptions of an orchestrating process provided by a service can be automatically created based on the functionalities of the services invoked by the orchestration.

1.3.9 QUALITY OF SERVICE

The quality of service (QoS) is the component of service metadata responsible for defining the qualitative properties (Chung 2000) that characterize services. Because service-oriented architectures are based on highly cooperative and open environments, the challenge is to capture the most relevant aspects, the corresponding metrics and how to measure QoS also taking into account that it could dynamically change. Some of the most prominent themes are service status, performance and security (for a complete survey see, e.g., Menasce 2002; Cardoso 2004). The service status describes information about the current state of the service: whether the service is deployed or it has been removed (or moved to a new location), whether it is currently available, and its current workload. The security is relevant because data and computation are distributed over the network and it becomes important, for instance, to formally check whether certain data can be considered as trustworthy or not.

Because the system could be equipped with a number of service instances supplying the same functionality (e.g. provided by different member states), QoS aspects play a fundamental role in the discovery where matching services can be filtered according to QoS requirements. For example, if different user profiles are considered, performance measures (e.g., average/maximum response time and maximum amount of clients supported) could be used to prioritize applications.

When orchestration is regarded, other problems become relevant for QoS, such as how to measure the quality of an entire orchestration, which depends on the QoS of all involved services. The QoS of orchestrated services additionally complicates the concrete definition of the workflows in the case that specific QoS requirements have to be fulfilled. Consider the case where one of the nonfunctional requirements is the support of a large quantity of concurrent clients; in order to guarantee a high level of scalability, the orchestrating service should be able to distribute the workload over multiple instances of services in a balanced way.

1.4   SUMMARY

In this chapter we described the background and the key objectives of a European Spatial Information Infrastructure, INSPIRE, and the complexity in building INSPIRE on top of national spatial information infrastructures established and operated by member states.

We have identified some important challenges that need to be addressed in the developments and implementations of INSPIRE. In particular, when we look at the pan-European scale, issues such as multilinguality and cross-border inconsistencies are immediate challenges that need to be addressed. Because INSPIRE anticipates the involvement of services in order to support distributed discovery, access and use of geographic information, this means moving the traditional GIS world towards a service-oriented architecture. In a service-oriented architecture additional challenges become critical, such as service functionality and quality of service. In the chapter, we have tried to provide the current status and guide future research in the areas that are important for the successful implementation for a European Spatial Information Infrastructure.

ACKNOWLEDGMENTS

We would like to thank our colleagues in the Spatial Data Infrastructures Unit for their input and help with this chapter. In particular, we would like to thank Nicole Ostländer and Michel Millot, who have given us valuable comments and suggestions.

REFERENCES

Afflerbach, S., A. Illert, and T. Sarjakoski. 2004. The Harmonisation Challenge of Core National Topographic Databases in the EU-Project GiMoDig. *XXth ISPRS Congress*, Istanbul, Turkey.

Baader, F., and W. Nutt. 2003. Basic Description Logics. In *The Description Logic Handbook. Theory, Implementation and Applications*, edited by F. Baader, D. Calvanese, D. McGuinness, D. Nardi, and P. Patel-Schneider. Cambridge: Cambridge University Press.

Battrick, B., ed. 2005. *GEOSS 10 Year Implementation Plan: Reference Document*, GEO 1000R / ESA SP-1284.

Benjamin, B., K. Knight, and D. Marcu. 2002. Translation by the Numbers: Language Weaver. *5th Conference of the Association for Machine Translation in the Americas* (AMTA 2002), October 8–12, 2002, at Tiburon, California.

Bernard, L., I. Kanellopoulos, A. Annoni, and P. Smits. 2005. The European Geoportal: One Step Towards the Establishment of a European Spatial Data Infrastructure. *Computers, Environment and Urban Systems* 29:15–31.

Bishr, Y. 1999. A Global Unique Persistent Object ID for Geospatial Information Sharing. *2nd International Conference on Interoperating Geographic Information Systems*, Zurich, Switzerland.

Brogi, A., C. Canal, and E. Pimentel. 2006. On the Semantics of Software Adaptation. *Science of Computer Programming* 61 (2):136–151.

Cardoso, J., A. Sheth, J. Miller, J. Arnold, and K. Kochut. 2004. Quality of Service for Workflows and Web Service Processes. *Journal of Web Semantics* 1 (3):281–308.

Chen, H., B. Schatz, T. Ng, J. Martinez, A. Kirchhoff, and C. Lin. 1996. A Parallel Computing Approach to Creating Engineering Concept Spaces for Semantic Retrieval: The Illinois Digital Library Initiative Project. *IEEE Transactions on Pattern Analysis and Machine Intelligence* 18 (8):771–782.

Chung, L., B. A. Nixon, E. Yu, and J. Mylopoulos. 2000. *Non-Functional Requirements in Software Engineering*. Dordrecht, The Netherlands: Kluwer Academic Publishing.

Commission of the European Communities. 1995. CORINE land cover. http://reports.eea. Europa.cu/CORO-landcover/en/land_cover.pdf (assessed January 31, 2008).

Commission of the European Communities. 2007. Directive 2007/2/EC of the European Parliament and of the Council of 14 March 2007 Establishing an Infrastructure for Spatial Information in the European Community (INSPIRE). Brussels: Commission of the European Communities.

Craglia, M., A. Annoni, M. Klopfer, C. Corbin, L. Hecht, G. Pichler, and P. Smits. 2003. Geographic Information in the Wider Europe. http://www.ec-gis.org/ginie/doc/ginie_book.pdf (accessed January 31, 2008).

Devogele, T., J. Trevisan, and L. Raynal. 1996. Building a Multi-Scale Database with Scale-Transition Relationships. *7th International Symposium on Spatial Data Handling*, Delft, The Netherlands.

Farrell, J. and H. Lausen, ed. 2007. Semantic Annotations for WSDL and XML Schema: W3C. Recommendation. http://www.w3.org/TR/sawsdl/ (accessed January 31, 2008).

Felluga, B., and P. Plini. 2004. *EDEN-Inland Waters TRS, Terminology Reference System*. Rome, Italy: Consiglio Nazionale delle Ricerche (CNR), Istituto sull'Inquinamento Atmospherico, Unità Terminologia Ambientale.

Fensel, D., H. Lausen, A. Polleres, J. D. Bruijn, M. Stollberg, D. Roman, and J. Domingue. 2007. *Enabling Semantic Web Services. The Web Service Modeling Ontology*. Berlin: Springer.

Friis-Christensen, A., M. Lutz, N. Ostländer, and L. Bernard. 2007. Designing Service Architectures for Distributed Geoprocessing: Challenges and Future Directions. *Transactions in GIS* 11(6):799–818.

Friis-Christensen, A., C. S. Jensen, J. P. Nytun, and D. Skogan. 2005. A Conceptual Schema Language for Managing Multiply Represented. *Transactions in GIS* 9 (3):345–380.

Furuti, C. A. 2003. *Useful Map Properties: Areas. Are Area Ratios Preserved?* October 2, http://www.progonos.com/furuti/MapProj/Normal/CartProp/AreaPres/areaPres.html (accessed July 18, 2007).

Gagnon, S. 2007. Special Issue on Service-Oriented Business Process Integration. *International Journal of Business Process Integration and Management* 2 (1).

Goldstone, R. L., and J. Son. 2005. Similarity. In *Cambridge Handbook of Thinking and Reasoning*, edited by K. Holyoak and R. Morrison. Cambridge: Cambridge University Press.

Goodchild, M. F., and S. Gopal. 1989. *The Accuracy of Spatial Databases*. London: Taylor & Francis.

Gratius, M., M. Bertran, and H. Rodriguez. 2004. Multilingual and Multimedia Information Retrieval from Web Documents. *15th International Workshop on Database and Expert Systems* Applications, Zaragoza, Spain.

Guptill, S. C., and J. L. Morrison. 1995. *Elements of Spatial Data Quality*. Amsterdam: Elsevier.

Hall, M. 2006. A Semantic Similarity Measure for Formal Ontologies (With an Application to Ontologies of a Geographic Kind). Fakultät für Wirtschaftswissenschaften und Informatik, Alpen-Adria Universität Klagenfurt, Klagenfurt, Austria.

Heuvelink, G. B. M. 1998. *Error Propagation in Environmental Modeling with GIS*. London: Taylor & Francis.

ISO/TC-211. 2002. ISO FDIS 19113 Geographic Information: Quality Principles. Geneva, Switzerland: International Organization for Standardization.

———. 2003a. ISO 19109 Geographic Information: Rules for Application Schema. Geneva, Switzerland: International Organization for Standardization.

———. 2003b. ISO 19114 Geographic Information: Quality Evaluation Procedure. Geneva, Switzerland: International Organization for Standardization.

———. 2003c. ISO 19115 Geographic Information: Metadata. Geneva, Switzerland: International Organization for Standardization.

Kashyap, V., and A. Sheth. 1996. Semantic and Schematic Similarities between Database Objects: A Context-Based Approach. *The VLDB Journal* 5 (4):276–304.

Kilpelainen, T. 1997. Multiple Representation and Generalization of Geo-Databases for Topographic Maps, Finnish Geodetic Institute.

Knight, K., and D. Marcu. 2005. Machine Translation in the Year 2004. *IEEE International Conference on Acoustics, Speech, and Signal Processing (ICASSP '05)*, 18–23 March 2005, Philadelphia, PA.

Leach, P. 2005. *A Universally Unique IDentifier (UUID) URN Namespace*. ftp://ftp.rfc-editor.org/in-notes/rfc4122.txt (accessed June 22, 2007).

Lemmens, R. 2006. Semantic Interoperability of Distributed Geo-Services. Ph.D. thesis, TU Delft, Delft, The Netherlands.

Lucke, S., P. Plini, V. De Santis, and B. Felluga. 2003. *EDEN-Inland Waters Glossary*. Rome, Italy: Consiglio Nazionale delle Ricerche (CNR), Istituto sull'Inquinamento Atmospherico, Unità Terminologia Ambientale.

Lutz, M., J. Witte, E. Klien, C. Schubert, and I. Christ. 2008. Overcoming Semantic Heterogeneity in Spatial Data Infrastructures. *Computers and Geosciences. Special Issue on Geoscience Knowledge Representation* (forthcoming).

Madhavan, J., P. A. Bernstein, and E. Rahm. 2001. Generic Schema Matching with Cupid. *27th International Conferences on Very Large Databases (VLDB)*, Rome, Italy.

Masser, I., ed. 2007. *Building European Spatial Data Infrastructures*. Redlands, CA: ESRI Press.

Menasce, D. A. 2002. QoS Issues in Web Services. *IEEE Internet Computing* 6 (6):72–75.

Miles, A., and D. Brickley. 2005. SKOS Core Guide. W3C Working Draft. http://www.w3.org/TR/swbp-skos-CORE-guide/ (accessed January 31, 2008).

Nebert, D. D. 2004. *Developing Spatial Data Infrastructures: The SDI Cookbook*, Version 2.0. http://www.gsdi.org/docs2004/cookbook/cookbookV2.0.pdf (accessed August 14, 2007).

Nogueras-Iso, J., M. Á. Latre, R. Bejar, P. R. Muro-Medrano, and F. J. Zarazaga-Soria. 2005. SDIGER: Experiences and Identification of Problems on the Creation of a Transnational SDI. Paper presented at Jornadas Técnicas de la Infraestructura de Datos Espaciales de España (JIDEE '05), 24–25 November 2005, Madrid, Spain.

Ordnance Survey. 2006. OS MasterMap Address Layer and Address Layer 2: User Guide. Ordnance Survey, UK.

Ostländer, N., S. Tegtmeyer, and T. Förster. 2005. Developing an SDI for Time-Variant and Multi-Lingual Information Dissemination and Data Distribution. Paper presented at 11th EC GI&GIS Workshop, ESDI: Setting the Framework, Alghero, Italy.

Patil, L., D. Dutta, and R. Sriram. 2005. Ontology-Based Exchange of Product Data Semantics. *IEEE Transactions on Automation Science and Engineering* 2 (3):213–225.

Phillips, D. L., and D. G. Marks. 1996. Spatial Uncertainty Analysis: Propagation of Interpolation Errors in Spatially Distributed Models. *Ecological Modelling* 91 (1–3):213–229.

Raubal, M. 2004. Formalizing Conceptual Spaces. In *Formal Ontology in Information Systems, Proceedings of the Third International Conference (FOIS 2004). Frontiers in Artificial Intelligence and Applications 114*, edited by A. Varzi and V. L. Amsterdam. Amsterdam: IOS Press.

Rodríguez, A., and M. Egenhofer. 2003. Determining Semantic Similarity Among Entity Classes from Different Ontologies. *IEEE Transactions on Knowledge and Data Engineering* 15 (2):442–456.

Sargent, P. 1999. Features Identities, Descriptors and Handles. *2nd International Conference on Interoperating Geographic Information Systems*, Zurich, Switzerland.

Senellart, J., P. Dienes, and T. Váradi. 2001. New Generation Systran Translation System. *MT Summit VIII, 18–22 September 2001*, Santiago de Compostela, Spain.

Sheth, A., S. Gala, and S. Navathe. 1989. Attribute Relationships: An Impediment in Automating Schema Integration. *NSF Workshop on Heterogeneous Database Systems*, Chicago, IL.

Shvaiko, P., and J. Euzenat. 2005. A Survey of Schema-Based Matching Approaches. *Journal of Data Semantics IV*, 146–171.

Sievers, J., and P.-G. Zaccheddu. 2005. EuroGeoNames: The Vision of Integrated Geographical Names Data within a European SDI. *Eighth United Nations Regional Cartographic Conference for the Americas*, 27 June–1 July 2005, New York.

Soergel, D., B. Lauser, A. Liang, F. Fisseha, J. Keizer, and S. Katz. 2004. Reengineering Thesauri for New Applications: The AGROVOC Example. *Journal of Digital Information* 4 (4).

Stuckenschmidt, H. 2003. Ontology-Based Information Sharing in Weakly Structured Environments. Ph.D. thesis, Vrije Universiteit, Amsterdam.

Tchistiakov, A., J. Jellema, H. Preuss, T. H. Diaz, B. Cannell, J. Passmore, T. Mardal, D. Capova, J. Belickas, and V. Rapsevicius. 2005. eEarth: Bridging the Divided National Geo-Databases via Multilingual Web Application. Paper presented at 10th International Symposium on Information and Communication Technologies in Urban and Spatial Planning and Impacts of ICT on Physical Space, Vienna University of Technology, 22–25 February, 2005, Vienna.

UNGIWG. 2007. *STRATEGY for Developing and Implementing a United Nations Spatial Data Infrastructure*. http://www.ungiwg.org/docs/unsdi/UNSDI_Strategy_Implementation_Paper.pdf (accessed August 2007).

van Oosterom, P. 2005. Variable-Scale Topological Data Structures Suitable for Progressive Data Transfer: The GAP-Face Tree and GAP-Edge Forest. *Cartography and Geographic Information Science* 32 (4):331–346.

Veregin, H. 1999. Data Quality Parameters. In *Geographical Information Systems: Principles and Technical Issues*, edited by P. A. Longley, M. F. Goodchild, D. J. Maguire and D. W. Rhind. New York: John Wiley & Sons.

Visser, U., and H. Stuckenschmidt. 2002. Interoperability in GIS: Enabling Technologies. Paper presented at 5th AGILE Conference on Geographic Information Science, Palma de Mallorca, Spain.

Visser, U., H. Stuckenschmidt, C. Schlieder, H. Wache, and I. Timm. 2002. Terminology Integration for the Management of Distributed Information Resources. *Künstliche Intelligenz* 16 (1):31–34.

Visser, U., T. Vögele, and C. Schlieder. 2002. Spatio-Terminological Information Retrieval Using the BUSTER System. Paper presented at Environmental Communication in the Information Society, *16th Conference on Informatics for Environmental Protection* (EnviroInfo), Vienna, Austria.

Vögele, T., and R. Spittel. 2004. Enhancing Spatial Data Infrastructures with Semantic Web Technologies. *7th Conference on Geographic Information Science (AGILE 2004)*, Heraklion, Greece.

W3C. 2006. *W3C: OWL Web Ontology Language Guide*. W3C 2004. http://www.w3.org/TR/2004/REC-owl-guide-20040210/ (accessed May 18, 2006).

Wache, H., T. Vögele, U. Visser, H. Stuckenschmidt, G. Schuster, H. Neumann, and S. Hübner. 2001. Ontology-Based Integration of Information: A Survey of Existing Approaches. Paper presented at IJCAI-01 Workshop Ontologies and Information Sharing, Seattle, WA.

Wang, R. Y., M. Ziad, and Y. W. Lee. 2001. *Data Quality*. Kluwer International Series on Advances in Database Systems. Dordrecht, The Neterlands: Kluwer Academic Publishers.

Zarazaga-Soria, F. J., J. Nogueras-Iso, M. A. Latre, A. Rodriguez, E. López, P. Vivas, and P. R. Muro-Medrano. 2006. Providing SDI Services in a Cross-Border Scenario: the SDIGER Project Use Case. Paper presented at GSDI-9 Conference, 6–10 November 2006, Santiago, Chile.

2 Geometry Semantics in Spatial Information

John R. Herring, Jayant Sharma,
Ravikanth V. Kothuri, and Siva Ravada

You do what you can, when you can, because you must.

Anonymous

CONTENTS

Coordinate geometry has always been central to spatial information because it can easily and accurately represent spatial extents. Requirements for spatial information are becoming more complex, and geometry is being used to represent some of that complexity, but in ways that may not be familiar to many GIS users. Not all spatial data are directly compatible with the more classical GIS application—CAD, architecture, engineering and construction (AEC)—and geospatial engineering data and its geometry have long been considered "different" and been barred from direct integration into geographic applications; spatial query, really a branch of spatial analysis, has been relegated in the past to a procedural application often requiring GIS expertise even to understand.

At the same time, geographic information requirements are being laid against end-user and public access applications outside the scope of classical GIS, such as architectural and engineering data in three dimensions. For this reason there is pressure to move geographic information into a service model with simple accesses to allow non-GIS application developers to integrate external geographic functionality into their applications without delving into the sea of geographic complexity. Although useful in many applications, this external service approach is limited and does not provide a good mechanism to integrate the diverse types of spatial data being used. Nongeographic applications with heavy spatial components, like engineering, need a deeper level of integration points into geographic space. Both deep technical and lighter public use are leading geographic information to a universality of use that will ultimately tend toward complete ubiquity, where spatial data will be on a par with textual and numeric data as the backbone of any technical endeavor. To be successful, an SII (spatial information infrastructure) must provide for both types of integration. But the paradoxical requirements for "ease of use," "public access" and "deep integration" are beyond the classical GIS approaches.

The successful implementation of applications to support these requirements may depend on new uses of abstract geometric techniques like topology, non-Euclidean metrics, projective and differential geometry in a wide range of applications. What was considered as "esoteric," "eclectic" or "pure mathematical disciplines" will become essential tools for everyone from the geographic-scale engineer to the casual user. These technologies all have "theoretical" characteristics that affect their value in "practical" issues from large-scale engineering projects to in-car navigation and other personal applications.

2.1   INTRODUCTION

The history of geographic information in the digital world has always been mainly a story of how geometry is handled. Within this shifting framework, this chapter examines the present and potential requirements affecting the spatial paradigm, with special interest in the integration of nongeographic spatial data into a spatial infrastructure. The core purpose of an SII is to create a single entry point for tasks involving spatial information. In addition to classical geographic data and their related fields, there are other spatial applications that have historically not been well-integrated. This has included engineering, architecture, graphics, simulation and other disciplines that tend to work in Euclidean coordinates, but nonetheless contain a significant treasury of useful data that could be shared with more classical geographic applications. The common point of contact is geometry. We just need to understand how their view differs from geography, and how we can both transform and integrate all the spatial disciplines while maintaining data independence between the various data and stakeholders of the systems.

2.2   GEOMETRY IS JUST ANOTHER
REPRESENTATION FOR INFORMATION

Geometry has always played an important role in geographic applications, but that role has changed as our understanding of data complexity has changed. The early geographic information applications, mimicking the paper process, used geometry viewed as graphics as a central organizing data type, and then applied attributes to describe what the geometry was representing. This "geometry-first" has fatal flaws, the two most important being the separation of spatial representation from other representational information, and the representation of information indirectly by *ad hoc* graphic symbolization. This made the information hard to use by anyone not knowing that "red curve on level 2, dashed with 2 longs and 1 short, was a road" or other secrets of the jargon-encrypted information.

The real break with this was the "feature model," often mislabeled the "object model." This looked at geometry as simply another attribute of some abstracted real-world entity. The mode of implication was reversed and something had a symbology because it had certain characteristics, not vice versa. This is currently the model for most GIS systems, although remnants of the old "geometry-first" paradigm still exist in aspects of some systems, such as the now-rare restriction to one spatial geometry value per feature.

The feature model is one key to the complexity issue. It is a simple vocabulary that is understandable by anyone who can read a map, and it is independent of the complexity of the coordinate geometry objects that are represented. Even the "ability to read a map" requirement for the user is fast disappearing—navigation systems that used to depend on map presentations are shifting towards virtual worlds where the display is shifting from maps to a simulation of ground-level visibility, a triumph of the "virtual window" over the map as the prevalent user-interface metaphor.

2.3   QUERY, INDEXING, DISTANCE, AND TOPOLOGY

Query in a spatial information database is the ability to find features based on their property values and relationships. Thus, query requires a consistent way to define and to discover the relationship between geometric representations. There are two categories of spatial relations: those that depend on the value of some measure (metric relations) and those that do not (topological relations). Metric relations are based on distance or direction. Most of our intuition is based on classical Euclidean geometry in a plane and will work for short distances. Spatial indexes such as r-trees and quad-trees can be used to filter distance queries as well as to locate "nearest" neighbors.

Bearing is another issue. Geodesics are generally not of constant "bearing," that is, their angle from north changes constantly. If we are speaking of the "bearing" from some source point towards some other target point, we have two obvious measures to choose from: (1) a geodesic bearing that is the exit angle of the best geodesic leaving the source point for the target point or (2) a "rhumb" line bearing—the bearing of a curve of constant bearing (a straight line on a Mercator projection) that goes from the source point to the target point.

For points fairly close to one another, the two are close to one another, but the longer the distance, the more they will tend to diverge. Extending the concept from points to larger extents also complicates the issue, and no consistent approach really exists.

2.3.1   TOPOLOGY, THE BASIS FOR NONMETRIC RELATIONS

Nonmetric relations that make sense must be invariant under deformations of space, that is, continuous transformations. These relations are "topological" in the sense that topology is the study of geometric properties invariant under such continuous transformations. The most powerful tool in the mathematical topologist toolbox is algebraic topology, where isomorphic (differing only by a continuous transformations) sets of geometric entities are symbolically manipulated using algebraic techniques. The details of how this is done are beyond the scope of this chapter, but the topic is well covered by mathematical texts; Spanier (1996) is a classic text using cellular complexes, but most graduate texts are similar. Alternatively, simplicial topology is a bit more computationally direct, and uses the multidimensional equivalent to a triangulated network; see Egenhofer et al. (1989).

The first move in geographic information from intuition to mathematics was from graphics to topology. This new view of geography as abstract data and not graphics opened the door to pure symbolic logic used for spatial analysis. Pictures helped explain what was going on but were no longer the core logic. Topology, the study of geometric properties that are invariant under transformations, operates by breaking geometry into simple, well-understood objects (such as convex simplices as defined above) and doing combinatorial analysis on these simple objects to determine how their composites interact. Although White (1984) was probably the first to see the potential of formal topology in dealing with geographic information, Egenhofer (1989) was the first to use it to define and investigate topological relations on geographic entities. It is interesting that most follow-up work to the original paper has been on using the classification method to "count and classify" relations, while the real power of the technique is the direct application of the computational method to database spatial query.

To determine how two geometric objects interact, they are put into a common topological structure, and a comparison is made on how each component of the space interacts. For each object A, one can calculate its interior iA, its boundary ∂A and its closure $cA = iA \cup \partial A$. For two objects A and B, that gives nine sets to look at (if the universal U is defined as the entire world) usually arranged in a 3×3 matrix (sometimes referred to as the "9-intersection method"):

$$\begin{bmatrix} iA \cap iB & \partial A \cap iB & iB - cA \\ iA \cap \partial B & \partial A \cap \partial B & \partial B - cA \\ iA - cB & \partial A - cB & U - (cA \cup cB) \end{bmatrix} \quad (2.1)$$

This is slightly different from the representation that is usually given, but it is more accurate as to the actual algorithm used. The usual representation uses the complement of A defined as $A^c = U - cA$, which is difficult to actually do because U is "big." The easier algorithm is to take the difference with the closure of A. The last entry $(U - (cA \cup cB))$ is almost always a nonempty set containing an element of all dimensions, and therefore almost never needs to be calculated. Each of the sets in the nine matrix is categorized as empty, containing a dimension-0 object and then up to containing a dimension-3 object.

"Error! Reference source not found". shows a simple example of two intersecting "area" features (the interiors of the circles). When looking at the nodes, edges and faces we have the topological representation of two circles C_{left} and C_{right} as lists of topological objects, as shown in the following:

$$cC_{right} = \{F1, F3, E1, E2, E3, N1, N2\}$$

$$\partial C_{right} = \{E1, E3, N1, N2\}$$

$$iC_{right} = \{F1, F3, E2\}$$

$$cC_{left} = \{F2, F3, E2, E3, E4, N1, N2\} \quad (2.2)$$

$$\partial C_{left} = \{E2, E4, N1, N2\}$$

$$iC_{left} = \{F2, F3, E3\}$$

This gives us the matrix of intersection:

$$\begin{bmatrix} iC_{right} \cap iC_{left} & \partial C_{right} \cap iC_{left} & iC_{left} - cC_{right} \\ iC_{right} \cap \partial C_{left} & \partial C_{right} \cap \partial C_{left} & \partial C_{left} - cC_{right} \\ iC_{right} - cC_{left} & \partial C_{right} - cC_{left} & U - (cC_{right} \cup cC_{left}) \end{bmatrix} = \begin{bmatrix} \{F3\} & \{E3\} & \{F2\} \\ \{E2\} & \{N1, N2\} & \{E4\} \\ \{F1\} & \{E1\} & \{F0\} \end{bmatrix} (2.3)$$

The last step in the query is to compare this matrix with a template, usually defined by the characteristics of the sets in equation 2.3. Such symbology may look like

$$0 = \text{empty set}$$

$$1 = \text{contains only nodes}$$

$$2 = \text{contains only nodes and edges} \quad (2.4)$$

$$3 = \text{contains only nodes, edges and faces}$$

$$* = \text{not restricted}$$

in which case the category of the relation is given by equation 2.5. Any other set of areas in the "general position" similar to those in figure 2.1 will have the same type of 9-intersection matrix.

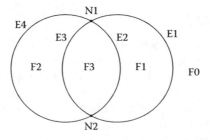

FIGURE 2.1 A simple topology structure.

$$\begin{bmatrix} \{F3\} & \{E3\} & \{F2\} \\ \{E2\} & \{N1,N2\} & \{E4\} \\ \{F1\} & \{E1\} & \{F0\} \end{bmatrix} \Rightarrow \begin{bmatrix} 3 & 2 & 3 \\ 2 & 1 & 2 \\ 3 & 2 & 3 \end{bmatrix} \qquad (2.5)$$

There are variations on this mechanism, usually changing the categories of equation 2.4. The more categories used, the finer the distinctions. Ultimately, equation 2.4 is bounded by the topology of the nine sets, that is, two example relations are equivalent if their configurations (as in figure 2.1) are topological equivalents. All such variations share the same general characters that once the topologies of the two objects have been calculated:

1. The process produces consistent results, and therefore standardization of the definitions of topology and the characterization of the matrix gives a deterministic result.
2. All the work after the calculation of the topology is susceptible to standard optimization techniques (indices, hash tables, symbolic sorts, etc.), yielding performance akin to nonspatial query.
3. Characterizations of topological relations (nonmetric) give query results that are intuitively correct and consistent. Although some such definitions can be based on naïve "common" language use, technical and legal standards may take precedence in many applications.
4. The addition of a few standard metric relations (distance, within distance, buffer zones, nearest neighbors, etc.), in addition to the generic topological ones, gives a "complete" spatial query language.

The only question left is one of optimization of topological calculations, which may depend more on who holds the data than the best approach, and a set of standardized matrix patterns for "named operators." Egenhofer (1989) set down a preliminary list, and the Open Geospatial Consortium (2007) has another. The key to the functionality in this approach is not that it allows theorists to categorize the topological relationships possible between different types of geometries, but that it allows data stores to use standard definitions for spatial relations that are unambiguously defined and that the method for calculating these relations is unified by the algorithms suggested by the matrix in equation 2.1.

2.4   CAD AND AEC

Engineering design and construction data are a source of detailed, accurate and reliable large-scale data available. But they represent a source of challenge for SII implementations, because they seldom meet the expectation for geographic information. CAD (computer aided design) or AEC (architecture, engineering and construction) applications are done in local Euclidean coordinate spaces, where all planes are flat, the Pythagorean theorem always works and all triangles have 180° worth of angles. All of this works fine, within reasonable limits of errors, as long as you are working within a few square miles or less, but the Earth is not flat, and eventually (at somewhere about

8 to 15 kilometers) it gets noticeable. Integrating this type of source into an SII presents special problems, most of which are related to this necessity to fix the coordinates, often when there is no geographic coordinate space involved. The usual mechanism is to "tie" a point in the design space to a point in a geographic coordinate space, and then use simple N-S, E-W and vertical offsets to locate other points. If you are dealing in earth-centric coordinate space, then you are essentially looking at a data shift (a little larger than is normally done) and the standard Bursa-Wolf data transformation can be used. This can be represented by seven parameters (three parallel offsets for an (x, y, z)-shift, three rotations about the various axes and one scale factor).

A similar formulation can be used locally for any spatial coordinate system. The first modification has to do with scale. On any map (a projection of a curved surface onto a flat plane) of any size, sooner or later the scale becomes direction dependent. Linearly fitting Euclid's idealized right-angled, flat coordinates to Eratosthenes' spherical Earth has its solutions in differential geometry (see Spivak 1999), and in particular the use of the concept of tangent spaces that associates to any point on a manifold a local Euclidean vector space based on directions. Similarly, the geometry of moving and morphing objects creates problems beyond static representation of classical GIS. For small enough radii, a linear approximation to a tangent plane to the Earth's surface is sufficient to obtain an approximation of the transformation from "local space rectangular" to "geographic coordinate space." A local tangent space on a manifold at a point is logically the vector space made up of all tangent vectors of all curves passing through the point. For an ellipsoidal 2D coordinate system, it would be best to define a local orthonormal frame for the horizontal direction, say $\vec{X}$ and $\vec{Y}$ that might be the north and south vectors, each normalized to a unit length using whatever measurement units are being used in the engineering design file, and then add a unit vertical $\vec{Z}$ that would always point upward (out from the center of the ellipsoid). This gives a reasonable local Euclidean frame for most points (there is a problem at the poles, but a topological theorem by Brouwer says that there will always be a vanishing point for a vector field tangent to a topological sphere, so this is unavoidable). The creation of unit vectors means that a local metric is used, since most geographic coordinates will not have a global metric (except geocentric systems, which are really just a really big 3D Euclidean space). Once a local frame is created, a mapping from a parametric Euclidean 3 space $E^3 = \{(u, v, w)\}$ to local geodesic coordinates at the point (x_0, y_0, z_0) would be given by

$$\begin{bmatrix} x \\ y \\ z \end{bmatrix} = \begin{bmatrix} \vec{X_0} & \vec{Y_0} & \vec{Z_0} \end{bmatrix} \begin{bmatrix} u \\ v \\ w \end{bmatrix} + \begin{bmatrix} x_0 \\ y_0 \\ z_0 \end{bmatrix}$$

where

$$\vec{X_0} = \vec{X}(x_0, y_0, z_0)$$

$$\vec{Y_0} = \vec{Y}(x_0, y_0, z_0)$$

$$\vec{Z_0} = \vec{Z}(x_0, y_0, z_0)$$

(2.6)

This mapping would be good for small distances where the curvature of the Earth did not "bend" the surface much away from the tangent plane at the point (x_0, y_0, z_0).* It should be noted that equation 2.6 is a Bursa-Wolf transformation if the three vectors in the matrix $(\overrightarrow{X_0}, \overrightarrow{Y_0}, \overrightarrow{Z_0})$ are orthogonal and of the same length. In such a case, the matrix is a scale and rotation transformation. When the local scales in the three cardinal dimensions are not equal, then the three vectors have different lengths.

Classical GIS usually makes the target for curve and surface functions as defined below in equation 2.9, a spatial reference system that would be E^2 or E^3, but our formulation allows other uses. We have already seen some uses for this functionality, in creating measures in the example discussed in section 2.6.1, figures 2.3 and 2.4, and in using time. Other examples include moving rigid objects where the function becomes $u \rightarrow (x, y, z, \theta_x, \theta_y, \theta_z, t)$, where θ_x, θ_y and θ_z are rotations about the primary axes in the local tangent frame defining the orientation of the moving object with respect to its design space with respect to a moving frame as described above, and t is time. This gives us a 7D target space $\vec{p} \otimes \vec{r} \otimes (t)$ of a position and rotation vectors, with an added temporal dimension. If the distributed frame is defined using a curve interpolation also, we could get a cure in a coordinate space of dimension $(2 \times 3 + 1 + 3 \times 3)$ or 16D:

$$\vec{p} \otimes \vec{r} \otimes (t) \otimes \vec{X} \otimes \vec{Y} \otimes \vec{Z} = (position) \otimes (rotation) \otimes (time) \otimes (frame) \quad (2.7)$$

2.5   MODELING STATIC AND MOVING SPATIAL OBJECTS ON THE EARTH'S SURFACE

Current geographic applications usually consider geometry as 2D or 3D, the classical map representation or the same thing on a "2½D" surface. This is a limited view, and the mathematical view of geometry as the graphs of functions opens up possibilities.

A major difference is the dimension of the coordinate space that the geometry targets. Static representations can be represented with 2D and 3D coordinates, but as soon as things move, the modeling gets complicated. First, time becomes a component of position (4D). The shape of the moving object is probably given once, in its own "body centric coordinate system," which must have a target frame in geographic space (three vectors of three dimensions each, or nine more dimensions, as seen in equation 2.6). Embedding the object in the target frame will require three rotations (the scaling is taken care of in the frame), which means three more dimensions. So, depending on the mechanism for representation, we may have 3 (space) + 1 (time) + 9 (local frame) + 3 (orientation) = 16   coordinate   offsets.   In figure 2.2, the spatial-temporal coordinate is

$$(x, y, z, t, \theta_x, \theta_y, \theta_z, x_0, x_1, x_2, y_0, y_1, y_2, z_0, z_1, z_2)$$

If the frame is given by a distributed frame on the underlying surface, then the object's spatial-temporal coordinate is

---

* Along great circle geodesics, measured vertical directions as compared to theoretical parallel lines vary about 32.3 to 32.4 arc seconds per kilometer depending on direction, or about 0.009°. At 2 km, a 100 m elevation would be off laterally by about 1 mm for this error in inclination. This is usually good enough for geographic information purposes.

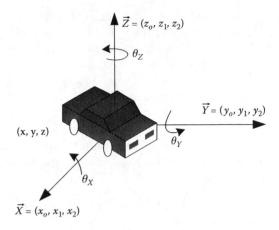

FIGURE 2.2   Nontemporal dimensions of a moving object.

$$(x, y, z, t, \theta_x, \theta_y, \theta_z) = position \otimes time \otimes rotations$$

and the underlying surface would have points (given as functions) like

$$(x, y, z, x_0, x_1, x_2, y_0, y_1, y_2, z_0, z_1, z_2) = position \otimes frame$$

More details are given in the example associated with equation 2.7.

Next, classical Euclidean geometries do not work, basically because the geographic coordinate systems are not Euclidean. Further, line strings and conics do not represent the spatial "shape" of many things when derivatives are involved.

2.5.1   3D Is Really 3D

In much of classical GIS, the third dimension is treated as a function of the first two, in essence drawing a map on a surface. This is not reality. In the real world, things at the same geographic position can be at different elevations, sometimes even the same thing. Things can often be classified as points, curves and surfaces, but the topological dimension (the thing that makes a curve in 3D coordinates different from a surface in 3D coordinates) is a characteristic of the object, not the space in which it is embedded.

The best way to express this concept is to take the concept of a "chart" from differential manifold theories and use it to define local geometries. In essence, geometry can be viewed as functions from a constructive parameter space into 3D coordinate space, something like

$$curve =: \{c : D \subset E^1 \rightarrow E^3\}$$

$$surface =: \{s : D \subset E^2 \rightarrow E^3\}$$

(2.8)

These functions have some constraints, mainly that they are locally invertible and bicontinuous (continuous as a mapping in either direction). The dimension of the parameter space is the topological dimension of the object. A curve, regardless of where it is embedded, is always an image of some part of the real line. Therefore, that curve can be defined as a function c from a contiguous subset of a 1D parameter space into a 3D geographic space, and a surface can be defined as a function s from a contiguous subset of a 2D parameter space to a 3D geographic space. ISO 19107 uses this type of definition to describe "curve segments" and "surface patches" that are to be combined along common boundaries so that mechanisms of interpolation can be changed or complex folds like tunnels and overpasses can be accommodated. This is essentially the equivalent of the concept of "charts" in manifold theory, where each point is surrounded by an appropriately dimensioned image of a parameter space.

Logic lets us extend this to zero, one, two, and three topological dimensions and to any dimension for the target space, but the interpretation feels odd to the nonmathematically inclined:

$$point =: \{p : E^o \rightarrow E^n \mid n \geq 0\}$$

$$curve =: \{c : D \subset E^1 \rightarrow E^n \mid n \geq 1\}$$

$$surface =: \{s : D \subset E^2 \rightarrow E^n \mid n \geq 2\} \qquad (2.9)$$

$$solid =: \{v : D \subset E^3 \rightarrow E^n \mid n \geq 3\}$$

The zero-dimensional parameter space is the "origin" and only the origin; so a point simply picks a single position in 3-space. A solid (v is for volume to prevent confusion with s for surface) uses what appears to be a self-embedding of 3-space into itself, but we must remember that the domain of these functions defines the topological dimension as local parameters and the range of the functions is probably a spatial (1D, 2D or 3D) or spatial temporal (2D, 3D, or 4D) coordinate system. That can change and the implementor can use the other dimensions to do a lot of really useful things.

2.5.2 TEMPORAL AS JUST ANOTHER DIMENSION

Once we have broken the connection between the surface parameter space (u, v) and the spatial coordinates (x, y, z), it takes very little to add coordinate dimensions to our types, say time, so our target coordinate space is (x, y, z, t) or $E^3 \otimes T$:

$$temporalPoint =: \{tp : E^o \rightarrow E^3 \otimes T\}$$

$$temporalCurve =: \{tc : D \subset E^1 \rightarrow E^3 \otimes T\}$$

$$temporalSurface =: \{ts : D \subset E^2 \rightarrow E^3 \otimes T\} \qquad (2.10)$$

$$temporalSolid =: \{tv : D \subset E^3 \rightarrow E^3 \otimes T\}$$

Now a temporal point is simply that, a point in time and space. If that needs to be a moving point, the underlying parameters need a temporal component, so we get moving objects, in which a temporal parameter dimension is equal to a temporal coordinate dimension:

$$movingPoint =: \{mp : D \subset E^0 \otimes T \to E^3 \otimes T\}$$

$$movingCurve =: \{mc : D \subset E^1 \otimes T \to E^3 \otimes T\}$$

$$movingSurface =: \{ms : D \subset E^2 \otimes T \to E^3 \otimes T\} \qquad (2.11)$$

$$movingSolid =: \{mv : D \subset E^3 \otimes T \to E^3 \otimes T\}$$

To maintain the semantics of the definition, the above functions need to preserve time, so that

$$\forall f \in \{mp, mc, ms, mv\} \Rightarrow \pi_t \circ f(u, v, w, t) = t$$

In other words, any of the "moving geometry" projection back in time must use the identity on time. This means that moving geometries in equation 2.11 can be viewed as "geometry-valued functions of time":

$$movingPoint =: \{mp : T \to \{point\}\}$$

$$movingCurve =: \{mc : T \to \{curve\}\}$$

$$movingSurface =: \{ms : T \to \{surface\}\} \qquad (2.12)$$

$$movingSolid =: \{mv : T \to \{solid\}\}$$

This gives us a mechanism other than "time slicing" to model temporal behavior or movable objects. A moving point is a single point whose location varies with time, and similarly for other dimensional objects.

The truly lovely thing about this is the mathematics for the normal objects, the temporal objects and the moving objects are all the same. This is because the equations for these types of function do not depend on the target dimension in most cases, but simply use vector-based arithmetic. For example, the equation for a line in a vector space between two given vectors is always the same, which is

$$\vec{L}(s) = (1-s)\vec{P_o} + s\vec{P_1}, \text{ where } s \in [0, 1]$$

As it turns out, this is true for a whole class of geometry (the approximating splines), the types of which share some very useful properties for the geographic application to depend on. This cuts the connection between viewing temporal databases as collections of time slices, and allows the representation of the history of a geometry-valued feature attribute as a mathematical function of time.

2.6   A DIGRESSION INTO APPROXIMATIONS

The usability of all of the mathematics above is dependent on pragmatic representations of functions that are both compact and efficient. The authors believe that two related types of function cover most cases: lines and splines.

2.6.1   PARAMETER-BASED AND COORDINATE-BASED FUNCTIONS

In looking at functions, there are a lot of numbers floating around, and it is important that their various uses are understood, before we look at how some numbers create others. Looking at a simple function such as $z = f(x, y)$ we get what is commonly but erroneously referred to as 2½D surfaces, where elevation is an expression of horizontal coordinates. On the other hand, a parametric surface function such as $(x, y, z) = F(u, v)$ is considered to be different, where, in truth, the first example is simply a version of the second, as can be seen by

$$(x, y, z) = F(x, y) = (x, y, f(x, y))$$

In the first example, one coordinate is represented as a function of other coordinates. In the more general case, all coordinates are represented as functions of some constructive parameters (u, v). This is really the same thing, where restrictions are added:

$$[(x, y, z) = (x, y, f(x, y))]$$

$$\Leftrightarrow \tag{2.13}$$

$$[x = u, y = v] \text{ and } [(x, y, z) = F(u, v) = (u, v, f(u, v))]$$

In some applications, parameters may be used to tie different functions together, so that if we have a 2D map view of a road, defined by a parametric function s as $(x, y) = (x(s), y(s)) = \vec{f}(s)$ (figure 2.3), and a cross-section elevation model defined by $z = g(s)$ (figure 2.4), we have a full 3D geometry defined by $(x, y, z) = (x(s), y(s), g(s)) = \vec{f}(s) \otimes g(s)$.

This separation of the horizontal map view from the vertical cross section is a common technique in road design and construction, and the final geometry of the road is often then expressed as a curve in 4D (x, y, z, s), where the parameter (the fourth dimension) may have been modified to be used as a linear referencing system to allow linkage to the road geometry by a 1D special coordinate system that maps s onto the full geometry $s \rightarrow (x, y, z)$. Another way to look at this it to remember that a function is a type of relation, and relations have an operation called the "equi-join." We can look at $f = (x, y, s)$ and $g = (z, s)$ and thus define $f \otimes_s g = (x, y, z, s)$.

In any case, a geometry value may be defined by interactions between various functions, which as graphs are in effect geometries of their own. Further, the limitation to 3D space or 4D space time may not be universal and other parameters (such as

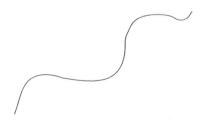

FIGURE 2.3   Two-dimensional road geometry defined by $(x, y) = f(s)$.

FIGURE 2.4 Elevation "cross-section" graph by "linear reference" $z = g(s)$.

"distance along the road" in our example above) may be carried as additional coordinate offsets, raising the dimension of the range of the function even further. This general technique is also valuable to interpolate separate offsets in the coordinate target space differently. For example, in a linear reference system, the parameter s is often in proportion to the length of the curve, and it may be more accurate to interpolate the curve, and then interpolate s based on a calculated arc length on the map.

2.6.2 LINEAR INTERPOLATION

The oldest form of interpolation used in geographic information is simple linear, the drawing of a straight line between two points. As observed before, if $s \in [0,1]$, then the function f on the closed interval $[0, 1]$ defined by $f(s) = (1-s)\vec{P_0} + s\vec{P_1}$ traces a line from the coordinate point $\vec{P_0}$ as its start to the coordinate point $\vec{P_1}$ as its end. By defining a set of real piecewise-line functions $L_n(s)$ using a graph as shown in figure 2.5, we can define a line string passing through a set of control points $\{\vec{P_n} \mid n = 0,1,2,...,N\}$ by

$$ls(u) = \sum_{n=0}^{N} L_n(u)\vec{P_n}$$

for $u \in [0,N]$, where

$$L_n(u) = \begin{cases} 0 & \text{if } u \notin [n-1,n+1] \\ u+1-n & \text{if } u \in [n-1,n] \\ n+1-u & \text{if } u \in [n,n+1] \end{cases}$$

(2.14)

Although not of any particular interest in its own right, this expression of a simple line string leads the way to a simple generalized definition for splines.

2.6.3 BEZIER, B-SPLINES AND NURBS

Spline functions come in two types. Fitting splines are curves that are forced to go through particular points and are usually found by solving sets of multivariate equations. Obviously a line string fits its control points (see equation 2.14), but it is not smooth. The most common fitted spline is a piecewise cubic polynomial that has a continuous first derivative. Getting smoother curves requires increasing the degree

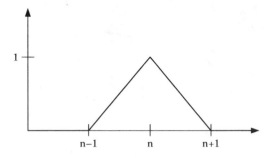

FIGURE 2.5   Graph of a linear weight function $L_n(s)$.

of the polynomial. Unfortunately, such splines are temperamental, and coordinate transformation will require a new series of solutions to their defining equation.

The form of equation 2.14 has the advantage that the formula remains the same as the control point array $\{\overrightarrow{P_n} \mid n = 0, 1, 2, ..., N\}$ changes. The use of other weight functions (higher-order polynomials) interferes with the fitting characteristics, but gives a set of geometry values that behave well under transformations. These are the approximating splines, and although they do not necessarily pass through their control points, they do act like smoothed versions of the line string passing through them. The most useful splines of this type are the Bézier (which use a set of polynomial weight functions) and B-splines (which use piecewise polynomial "basis" functions as weights). Both types have equivalent rational versions (polynomials are replaced by "rational" functions, which are quotients of polynomials). Beziers have a "global control" and every control point affects every value to some extent, whereas B-splines have a "local control" in that only a few points (one larger than the degree of the polynomials being used) affect any value along the curve. It turns out that B-splines are equivalent to piecewise Bézier curves, in that each section of a B-spline can be recast (with a derivable set of controls) as a Bézier spline. Similarly, a rational B-spline can be recast piecewise to a set of rational Beziers. This is an eclectic bit of information, but it is useful to know, because the same algorithms that do this transformation can be used to insert a "break point" that allows, without changing the geometry, one to divide a single spline defined geometry value into two. Several books on splines are listed in the bibliography (Bartels, Farin, and Rogers), and the Web has a whole set of valuable, and mostly reliable, resources for both information and open source code for splines (http://mathworld.wolfram.com/, http://planetmath.org/, http://www.geom.uiuc.edu/, http://www.mathwright.com/).

2.7   CONVEX GEOMETRY CONSTRUCTIONS

Triangles and triangulated networks have always been a mainstay of geographic information. They are an example of a fairly simple idea on convexity. A convex geometry is one in which a line between any two points in the geometry is wholly contained within the geometry. In two dimensions triangles are the simplest, but not the only, convex polygons. They even come with their own built-in interpolation mechanism—barycentric coordinates. Given three points,

$$\{\vec{P_i} \mid i = 0,1,2\}$$

and three non-negative numbers,

$$\{a_n \mid i = 0,1,2\} \text{ with } \sum_{i=0}^{2} a_i = 1$$

then the vector sum

$$\vec{P} = \sum_{i=0}^{2} a_i \vec{P_i}$$

is a point in or on the triangle formed by the original three points (if the points are colinear, this collapses to a line segment). Further, if we augment each of the points $\vec{P_i}$ with a parameter value u_i, then the resulting sum

$$(\vec{P} \otimes (u)) = Q = \sum_{i=0}^{2} a_i \vec{Q_i} = \sum_{i=0}^{2} a_i (\vec{P_i}, u_i)$$

is a linear interpolation of the parameter u at the point $\vec{P}$.

Given a triangulated irregular network (a set of points organized to form disjoint triangles), then any parameter (either numeric or vector) can be linearly interpolated using the barycentric coordinates in this fashion. In going to 3D objects in 3D space, the same works, except that instead of triangles, with up to four noncoplaner points, we get tetrahedrons. So, in general:

$$\text{Given } \{\vec{P_i} \in E^n \mid i = 0,1,2,...,m < n\}$$

where $\{\vec{P_i} \text{-} \vec{P_0} \mid i = 1,...,m\}$ are linearly indepdendent as vectors, then the set

$$\{P = \sum_{i=0}^{m} a_i \vec{P_i} \mid 0 \le a_i \le 1, \sum_{i=0}^{m} a_i = 1\}$$

(2.15)

is a convex geometry with barycentric coordinates $\{a_i\}$

The object so defined is called an m-simplex, and is fundamental to some forms of computational topology. In our spatial world, we will normally only deal with triangles and tetrahedrons, and their network. In all cases, the barycentric coordinate interpolation gives us a way to interpolate either scalar- or vector-valued fields (that is, with a scalar multiplication and an addition operation) on all points within the complex or on its boundary.

Again, since the interpolation is essentially linear, transformations, although not perfect, behave well under coordinate transformations, and in moving one of these networks, it is usually sufficient to move the defining points at the corners of the simplices.

2.8   MAKING SIMPLE USE OF COMPLEX FUNCTIONALITY

The areas covered so far in this chapter have made things more complex but more powerful. In this section we cover topics that make that complex functionality simpler to use.

2.8.1   FEATURE MODEL

A feature model simply exposes all the information in a database as a set of features, which are digital representations of real-world things (see ISO 19109). Query and searching are all cast as looking for features (usually of a particular type) with particular property values (see SQL/MM Spatial for an explanation of how to do this with SQL). Creating and editing are feature based, and issues as to the type of geometry to be used are all determined by the feature schema, with the user interface prompting for particular actions as appropriate for the type of feature or type of property being manipulated.

2.8.2   COVERAGE MODEL AND SPATIAL INTERPOLATIONS

The issue of deciding between raster and vector is also hidden by the feature model. An image or "coverage" is simply another property type for a feature, and the same context-sensitive manipulation can guide users through editing or querying by the type of real-world feature being sought, even if that is a satellite image. Furthermore, in terms of functions, an image is a surface mapping row and column into spectral values, so that $image =: \{s : D \subset E^2 \to E^n \mid n \geq 2\}$, where the dimension of the range is dependent on the number of spectral bands that the image has. If the image mapping includes a geometric "georeference" for the pixels (assuming three spectral bands for red, green and blue), then the mapping is $(u, v) \to (x, y, r, g, b)$.* The point here is that images are geometry.

2.9   CONSISTENCY ISSUES

Both technical and public use of spatial information is leading geographic information to a universality of use that will ultimately tend toward complete ubiquity, where spatial data will be on a par with textual and numeric data as the backbone of any technical endeavor. So, while the technology is becoming more complex, the requirements for a universal user interface are forcing implementations to simplicity and to automation. The paradox is that this simplicity of user interface and ubiquity of use requires a more complex infrastructure (SII and software to support it) than exists today, but a simpler interface. The average, public user is not interested in data sets, and he is not going to spend a lot of time manipulating data to derive some deep analytic result. He is interested in a few feature instances, and the answers to a few relatively simple questions about those instances, and an image or two to help understand that answer.

---

* The interpolation between grid points is often a cubic convolution, which is a type of spline function.

The root problem is the consistency of multiple data sources. The realities of the real world will prevent anyone from creating a universal data representation and data store. Issues preventing this may be social, political, legal (data ownership) and even technical. Problems such as multiple use, multiple coordinate systems and multiple scales may result in radically different data requirements. So the community will have to decide on and standardize the processes needed to make multiple data sources compatible and consistent. For geometric data conflation, in addition to the obvious solutions such as increased accuracy of data collection, there are practical solutions involving the comparison of topological structures, and the use of multidimensional splines to define coordinate-fitting transformations.

The major requirement driving this need is the integrity and consistency of answers. If different applications ask related questions about an area, the requirement should be that all the answers are consistent with one another. This may require more of a political and social solution than a technical one.

2.10   SUMMARY

In summary, the requirements placed on the geometric representations of features from multiple independent sources, in combination with the movement of geographic information to ubiquity, is paradoxically driving geographic information applications to greater technical complexity in geometric object implementation and greater simplicity in user interface. Part of that complexity and its solution is in the use of the mathematics or "functional representations" for geometry. The same interpolation mechanisms that can mold parameters to the shape of real-world features can be used to express the numeric relationships between space, time, distance and direction, and in the case of imagery, spectral values.

Viewing geometry values not solely in spatial terms, but in terms of functions from and to coordinate domains, allows users of spatial data to store and interpret different types of spatial and spatially related data using a consistent set of tools. Use of differential geometry, topology, and other deeper mathematical tools enables an SII to integrate geometric-based applications into a single spatial framework. Use of a universal concept of the feature model allows a simple and easy-to-understand user interface. Use of topologically defined spatial relations allows for simple, efficient, and unambiguous implementations of spatial query and analysis.

The geographic information community cannot always control the way in which spatial information is used, but it can build an environment of standards and best engineering practices that unifies the field and influences the manner in which such data are accessed, visualized, manipulated, and queried by applications that the public will eventually use.

REFERENCES

Bartels, Richard A., John C. Beatty, and Brian A. Barsky. 1987. *An Introduction to Splines for Use in Computer Graphics and Geometric Modeling.* Los Altos, CA: Morgan Kaufman Publishers.

Egenhofer, Max J. 1989. A Formal Definition of Binary Topological Relations. In *3rd International Conference on Foundations of Data Organization and Algorithms (FODO '89),* June 21–23, 1989, Paris, 457–472.

Egenhofer, Max, A. U. Frank, and J. Jackson. 1989. A Topological Data Model for Spatial Databases. In *Design and Implementation of Large Spatial Databases, Proceedings, First Symposium SSD '89*, Santa Barbara, California, July 1989. Lecture Notes in Computer Science, no. 409, ed. A. Buchmann, O Günther, T. R. Smith, and Y. F. Wang, 271–286.

Farin, G. 1999. NURB Curves and Surfaces from Projective Geometry to Practical Use, 2nd ed. Wellesley, MA: A. K. Peters Ltd.

ISO 19101. Geographic Informati—Reference Model.*

ISO 19107. Geographic Information—Spatial Schema.

ISO 19109. Geographic Information—Rules for Application Schema.

ISO 19133. Geographic Information—Location Based Services—Tracking and Navigation.

ISO/IEC 13249-3. Information Technology—Database Languages—SQL Multimedia and Application Packages—Part 3: Spatial.

ISO/IEC 9075. Information Technology—Database Languages—SQL.

Open Geospatial Consortium. 2007. OpenGIS® Implementation Specification for Geographic Information—Simple Feature Access—Part 1: Common Architecture, v1.2, available at http://www.opengeospatial.org/standards/sfa

Rogers, D. F., and J. A. Adams. 1990. *Mathematical Elements for Computer Graphics*, 2nd ed. New York: McGraw-Hill.

Spanier, E. H. 1966. *Algebraic Topology.* New York: Springer-Verlag.

Spivak, M. 1999. *A Comprehensive Introduction to Differential Geometry*, 3rd ed.,Vols. 1–5. Publish or Perish, Inc.

White, M. S. 1984. Technical Requirements and Standards for a Multipurpose Geographic Data System. *American Cartographer*, Washington, DC: American Congress of Surveying and Mapping, 11 (1):15–26.

---

* ISO standards are undated since the latest version is always the most useful. Information on ISO standards can be obtained at http://www.ios.org/.

3 Semantic Web Technologies as the Foundation for the Information Infrastructure

Frank van Harmelen

CONTENTS

The Semantic Web has emerged over the past few years as a realistic option for a worldwide information infrastructure, with its promises of semantic interoperability and serendipitous reuse. In this chapter we analyze the essential ingredients of semantic technologies, what makes them suitable as the foundation for the information infrastructure, and what the alternatives to semantic technologies would be as foundations for the information infrastructure. We will survey the most important achievements on semantic technologies in the past few years, and point to the most important challenges that remain to be solved.

3.1   HISTORICAL TREND TOWARDS INCREASING DEMANDS ON INTEROPERABILITY

When Thomas Watson, the founder of IBM, was asked for his estimate of how many computers would be needed worldwide, his reply is widely claimed to have been: "about five."* Of course, this presumed reply was given in 1943, but it shows the enormous shift in perspective that has taken place since the very early days of computing. Right until the late 1970s, the dominant perspective on computing was that of mainframe computing: large machines that provided centralized means of computing and data storage. In such a centralized perspective, interoperability is not the main concern: data are locked up in a centralized location, movement of data is rare, and if data are to be integrated, a special-purpose ad hoc transformation procedure is applied to transform the data into the required format.

The first revolution that was a major upset to the centralized perspective was the advent of the personal computer (PC) in the 1980s (ironically enough, also dominated by IBM). Suddenly, there were millions of small computing devices, each of which was capable of storing its own data, without recourse to centralized data storage. In this context, interoperability of data was becoming a problem: how to combine the data set stored in (or generated on) one PC with that of another PC, where another user, in a different organization, would be generating his or her own data?

However, the low degree of connectivity between the different PCs still kept the interoperability problem at bay. It was only the second revolution that really caused the data interoperability problem to bite, namely, the advent of the Internet (also arising in the 1980s), culminating in the rapid growth of the World Wide Web in the 1990s. The Internet has solved most wide-area networking problems with its nearly universally supported TCP/IP Internet Protocol and its DNS (Domain Name System) host-addressing scheme.

Suddenly, it became possible to exchange information from any computer to any other computer, and between any two users on the planet. In such a setting, special-purpose and ad hoc transformation procedures to import data are no longer a feasible alternative, and more principled mechanisms to ensure interoperability are required.

3.2   INTEROPERABILITY AT DIFFERENT ABSTRACTION LAYERS

The problem of interoperability in any information infrastructure (be it worldwide or local, be it for geographic information or otherwise) can be analyzed at different layers of abstraction (also see chapter 8, figure 8.1), all of which must be solved in order to obtain full interoperability:

Physical interoperability concerns the lowest layer of the abstraction hierarchy—plug-shapes and sizes, voltages, frequencies, and the bottom layers of the ISO/OSI network hierarchy. This is where most of the progress has been made, and physical interoperability between systems has been

---

* Although this quote is widely questioned now (http://en.wikipedia.org/wiki/Thomas_J._Watson), it makes the point.

solved; with the advent of hardware standards such as Ethernet, and with protocols such as TCP/IP and HTTP (Hypertext Transfer Protocol), we can nowadays walk into someone else's house or office and successfully plug our computer into the network (even automatically via wireless LANs), giving instant worldwide physical connectivity. Ironically, it is the success with which the physical interoperability problem has been solved that now creates problems at higher interoperability levels.

Syntactic interoperability—physical connectivity is not sufficient. We must also agree on the *syntactic form* of the messages we will exchange. Again, much progress has been made in recent years, particularly with the advent of eXtendible Markup Language, XML. XML has been dubbed "the ASCII of the 21st century," and indeed it is now the most widely used syntactic standard, and is itself used as a carrier for other syntactic standards such as HTML (for the content of Web pages*), WSDL (Web Service Description Language,† for the interfaces of Web services), and SOAP (Simple Object Access Protocol,‡ for the format of Web service messages).

Semantic interoperability—of course, even syntactic interoperability is not enough. We need not only agree on the form of the messages we exchange (structure of the information), or the form of the Web pages that we publish, but we also need to know the intended meaning of such messages and pages. In case we also want machine processing, for example, in urgent situations where human decisions have to be supported by machine processing (selections, combinations, translations, and other reasoning tasks), then the intended meaning has to be formalized.

3.3   THE MEANING OF SEMANTIC INTEROPERABILITY

In this section we shall be somewhat more precise about the meaning of semantic interoperability. Semantic interoperability is usually defined in terms of a formal semantics, and this can be done either denotationally, inferentially, or operationally. Although the primary definition of the semantics of formal languages is most often in terms of a denotational semantics (e.g., Hayes 2004; Patel-Schneider et al. 2004), for RDF (the Resource Description Framework) and OWL (the Web Ontology Language), respectively, we will instead describe semantic interoperability in terms of inferential semantics.

When an agent (a Web server, a Web service, a database, or a human in a dialogue) utters a message, the message will often contain more meaning than only the tokens that are explicitly present in the message itself. Instead, when uttering the message, the agent has in mind a number of "unspoken," implicit consequences of that message. When a Web page contains the message "Amsterdam is the capital of The Netherlands," then one of the unspoken, implicit consequences of this is that Amsterdam is apparently a city (because capitals are cities), that The Hague is

---

* http://www.w3.org/MarkUp/.
† http://www.w3.org/TR/wsdl.
‡ http://www.w3.org/TR/soap/.

not the capital of The Netherlands (because every country only has precisely one capital), etc. If agent A utters the statement about Amsterdam to agent B, they can only be said to be truly semantically interoperating if B not only knows the literal content of the phrase uttered by A, but also understands a multitude of implicit consequences of that statement, which are then shared by A and B.

3.3.1 MINIMAL SEMANTIC INTEROPERABILITY: EXPLICIT CONTENT ONLY

Thus, we could say that the semantic interoperability between A and B increases with the amount of information that they agree on after having exchanged a message. The minimal amount of information that they share is only the fact expressed in the statement itself: there is some object "Amsterdam" and some object "The Netherlands," and they are related by the first "being the capital of" the second. Notice that this minimal amount of semantic interoperability is already nontrival. Simply exchanging the following arbitrary XML syntax:

```
<is-capital-of>
   <Amsterdam/>
   <Netherlands/>
</is-capital-of>
```

is by itself *not* enough for B to understand that we are dealing with two objects and a relation between them: is "being capital of" a relation between two objects (as indeed intended), or does the tree-structure of the XML denote some-type information, as in

```
<humans>
   <males/>
   <females/>
</humans>
```

or does it denote some part-of information, as in

```
<heart>
   <left-chamber/>
   <right-chamber/>
<heart>
```

or any of an infinite number of other plausible semantic interpretations of the same syntactic structure? Thus, even to obtain from the earlier XML

```
<is-capital-of>
   <Amsterdam/>
   <Netherlands/>
</is-capital-of>
```

the minimal intended meaning that we are dealing with a relation between two objects, agents A and B must have previously agreed on this intended meaning of their syntactic structure, namely, that the root of the XML tree is the relation between a subject (first subnode) and an object (second subnode). In the context of the semantic Web, this is exactly the amount of semantic interoperability that RDF enables (that is, RDF without RDF schema). It allows us to pass single sentences, and only the literal content of those sentences themselves are guaranteed to be shared with any other agent adhering to the RDF semantics. Of course, the precise syntactic encoding is arbitrary (as long as it is agreed upon), and we could have written the above as

```
<relation name='capital-of'>
   <object name='Amsterdam'/>
   <object name='The Netherlands'/>
</relation>
```

or indeed as

```
<rdf:Description rdf:about='3116'>
   <name>Amsterdam</name>
   <isCapitalOf>The Netherlands</isCapitalOf>
</rdf:Description>
```

as it would read in RDF syntax.

3.3.2 EXTENDED SEMANTIC INTEROPERABILITY: SHARED INFERENCES

Although obtaining even minimal semantic interoperability from purely syntactic structures is nontrivial, it is of course a very limited form of semantics. In any reasonable human conversation, saying that "Amsterdam is the capital of The Netherlands" would also imply a number of other, unspoken facts, implicitly implied by what was said: that Amsterdam is apparently a city (because capitals are cities), that The Hague is not the capital of The Netherlands (because every country only has precisely one capital), and that The Netherlands is a country, or a province, but not another city, because countries and provinces have capitals, but cities do not. A spatial implied fact is the location of the capital city is inside the area covered by the country.

Thus, a more extended form of semantic interoperability would guarantee that if agent A utters a sentence S to B, then not only does B believe the literal contents of S, but B should also believe a number of other facts that can be inferred from S in combination with shared knowledge between A and B. It is exactly this shared knowledge that has become known as the shared *ontology* between A and B. In our little example, if such an ontology would indeed capture the fact that capitals are cities, capitals are unique, countries have capitals, etc., then A and B are guaranteed to have a much better basis for exchanging the intended meaning (semantics) of sentence S beyond its limited literal content. In fact, we could say that the amount of semantic interoperability between A and B is measured by the number of new facts that they both subscribe to after having exchanged a given sentence: the larger and richer their shared ontology, the more semantically interoperable they are.

It is exactly this kind of shared ontological information that can be captured in RDF Schema (as opposed to RDF only):*

```
<rdfs:Class rdf:about='Capital'>
    <rdfs:subClassOf rdf:resource='#City'/>
</rdfs:Class>
<rdf:Property rdf:ID='isCapitalOf'>
    <rdfs:domain rdf:resource='#Capital'/>
    <rdfs:range rdf:resource='#Country'/>
</rdf:Property>
```

states that capitals are cities, and that capitals are capitals of countries (allowing one to infer that Amsterdam must be a city if it is the capital of The Netherlands).

A more expressive language such as OWL is required to express that the capital of a country is unique:

```
<owl:InverseFunctionalProperty rdf:ID='isCapitalOf'/>
```

(the semantics of InverseFunctionalProperty states that the value of such a property uniquely defines the object of the property, because the inverse property (from value to object) is functional (has exactly one value)).

3.3.3 FULL SEMANTIC INTEROPERABILITY: UPPER AND LOWER BOUNDS

OWL is more expressive than RDF Schema in a very specific way: when agreeing on an ontology O expressed in RDF Schema, two agents A and B have both committed to a *minimal* set of beliefs that they will both uphold given some sentences S to be exchanged in the future; again, in our example, if A states that Amsterdam is the capital of The Netherlands, then by subscribing to their shared ontology O, B is forced to believe a number of other things as well (Amsterdam being a city, etc.). Hence, there is a minimum set of beliefs, a *lower bound*, on what agents A and B will infer after having exchanged a sentence S.

However, using an RDF Schema ontology, A cannot *forbid* B to believe certain things, for example, it cannot forbid B to believe that besides Amsterdam, The Hague is also a Dutch capital. Technically, this amounts to saying that RDF Schema cannot express negative information, because it does not contain negation. OWL does, and hence in OWL we can say that if Amsterdam is the Dutch capital, no other city can be (the InverseFunctionalProperty above). OWL enables A and B to not only put a lower bound on what they must believe after exchanging a sentence, it also allows them to put an *upper bound* on what they may not believe after exchanging a sentence. By strengthening the ontology O, A and B can move these lower and upper bounds successively closer together, hence narrowing the window of opportunity for

---

* #City is a shorthand URI referring to the concept City, defined at the same location as where the above statement can be found.

any misunderstandings (consisting of things that one of them believes after S, while the other one does not). Stronger ontologies, which place higher lower bounds and lower upper bounds on the set of inferred consequences of an exchanged sentence, increase the semantic interoperability between two agents.

3.3.4 THE ROLE OF XML AS A NOTATION

The above semantic descriptions have all been given in the XML notation that is prescribed by the W3C standards (Bechhofer et al. 2004 for OWL and Becket 2004 for RDF). However, this is only *one* particular syntax in which we can state the intended semantics. Of course, the XML syntax has a special status, as it is the one that was chosen for the standardization of documents. Nevertheless, different syntactic forms exist for expressing the same semantic contents. Examples in the case of RDF and OWL are the N3 syntax* that is popular because it is much more compact and readable than the official XML syntax. Also very popular is the UML-based notation because of its graphical presentation. In fact, many of the chapters in this volume (e.g. chapters 3, 4, 5, and 9) use UML class-diagrams to present knowledge that is also easily formalizable in OWL's official XML syntax.

3.4 THE SEMANTIC WEB AS A FOUNDATION FOR AN INFORMATION INFRASTRUCTURE

We are now in a position to define the role of Semantic Web technology as a foundation for an information infrastructure. The Semantic Web offers technology that contributes towards solving the interoperability problem at all three of the layers discussed above: HTTP, DNS, and URIs for physical interoperability; XML for syntactic interoperability; and RDF, RDF Schema, and OWL for semantic interoperability.

These languages and their corresponding technology are organized in a stack, where each higher layer uses lower ones. Semantic Web applications achieve semantic interoperability not only by exchanging their data, but also by exchanging (or having previously agreed to) explicit models of these data. These shared data models are often known as ontologies, and constitute shared knowledge used to interpret the information to be exchanged.

The most important premise on which the Semantic Web rests can be now be phrased as follows:

Premise: In order to achieve semantic interoperability it is certainly necessary (and most likely also sufficient) to express both data and data models (also known as ontologies) in languages with a formalized semantics, which enforces the sets of beliefs that agents must or may not uphold as the result of exchanging a certain piece of information.

Notice that this premise is very close to the knowledge representation hypothesis, formulated by Brian Smith in his 1982 Ph.D. thesis (Smith 1982, 15):

---

* http://www.w3.org/2000/10/swap/Primer.

Any mechanically embodied intelligent process will be comprised of structural ingredients that a) we as external observers naturally take to represent a propositional account of the knowledge that the overall process exhibits, and b) independent of such external semantical attribution, play a formal but causal and essential role in engendering the behavior that manifests that knowledge.

We can recognize "the propositional account of knowledge" in the propositional structure of the Semantic Web languages (RDF, RDF Schema, OWL), and the "causal and essential role in engendering behavior" is similar to the inference-enforcing process based on shared background knowledge that we discussed in the previous section.

We should note that although this approach to semantic interoperability may sound plausible, it is certainly not the only possible route. In particular, the possibility of a propositional account of the required knowledge to express sufficiently rich data models has traditionally been criticized in knowledge representation, and similarly in the Semantic Web. And it must be acknowledged that the currently most effective approaches to search are not based on *propositional* accounts of the contents of the Web, but rather on *statistical* models, as used in search engines such as Google, using word frequencies and patterns of links between pages instead.

Although the two approaches (the propositional and the statistical) are often positioned as alternatives, there is in fact nothing to make them mutually exclusive. It is quite possible, for example, to imagine statistical patterns being used as the basis for constructing a propositional account, and in fact many machine-learning contributions to Semantic Web technology (e.g., ontology learning) take exactly this combined approach.

3.5   MOST IMPORTANT ACHIEVEMENTS TO DATE

After having outlined the foundational ideas underlying the Semantic Web, and having described their role in a semantically interoperable information infrastructure, in this section we will discuss the most important achievements in recent years towards the realization of this Semantic Web Information Architecture.

3.5.1   ONTOLOGY LANGUAGES

A crucial and widely known achievement has been the definition and adoption of a number of data-modelling languages, stacked one on top of the other, with ever more expressivity: RDF, RDF Schema, and OWL, where the latter is itself divided into three substrata with proper syntax and semantic inclusions. Without going into the details of these languages (ample reference and teaching material exists), their general power is as follows.

- *RDF*: expressing binary relations between objects, and expressing that an object belongs to a given type (or class)
- *RDF Schema*: arranging these classes and properties in a class and property inheritance hierarchy (superclass-subclasss), and stating that properties have certain types as their domain and range

- *OWL Lite*: expressing (in)equalities between individuals, between classes, and between properties, and stating algebraic properties of properties (transitivity, symmetry, inverse functionality, etc), 0/1 restrictions on the cardinality, and ranges of properties
- *OWL DL*: definition of classes by enumeration, algebraic operations on classes (intersection, union, complement), stating disjointedness of classes, arbitrary cardinality restrictions on properties
- *OWL Full* introduces no new language constructions, but is more liberal in the way these constructions are combined (for example, using classes as instances of other classes)

Of course the increase in expressivity comes with an increase in computational costs of doing inference in these languages. The above stack of languages allows users to pick the language with the appropriate cost/benefit trade-off for each particular application. The (almost) proper inclusion relations between these languages ensure the possibility to move to more expressive languages as and when the need arises, without having to redo prior efforts.

With their status of W3C recommendation, these languages are guaranteed to be implementable and stable, enabling a rapid growth of industrial investment in their support and deployment.

3.5.2 ONTOLOGY VOCABULARIES

The above Semantic Web data-modelling languages have indeed been used for the construction of data models in a wide variety of domains. Very often, this was not a construction from scratch, but instead involved translating previously existing structures (data models, ontologies, thesauri, vocabularies) into these new languages with their standardized syntax and semantics, thereby illustrating how these languages do indeed enable greater degrees of semantic interoperability.

It is already impossible to be exhaustive (the Swoogly Semantic Web search engine lists more than 10,000 different vocabularies at the time of writing). We give the following incomplete list only to illustrate the diversity of ontologies expressed in Semantic Web languages, both in choice of domain and in how extensive the modelling has been.

- Biomedical: GO (15,000 terms from molecular biology*), SNOMED (300,000 terms from general medicine†), UMLS (a loosely integrated collection of over 100 medical vocabularies‡), FMA (205,000 concepts describing anatomy§).
- Top-level: Cyc (hundreds of thousands of terms and millions of assertions capturing commonsense knowledge¶), SUMO (20,000 terms and 60,000

---

* http://www.geneontology.org/.
† http://www.snomed.org/.
‡ http://umlsinfo.nlm.nih.gov/.
§ http://fma.biostr.washington.edu/.
¶ http://www.cyc.com/.

axioms capturing commonsense knowledge plus various specialized domains*), WordNet (115,000 definitions for about 150,000 words from the English language†).

- Cultural heritage: AAT (34,000 concepts and 131,000 terms relating to fine art, architecture, decorative arts, archival materials, and material culture‡), IconClass (28,000 definitions of objects, persons, events, and abstract ideas that can be the subject of an image§), ULAN (293,000 names and biographical and bibliographic information about artists and architects¶).
- Geographical: see chapter 8 for a description of a number of geographical vocabularies and ontologies.

Of course, these languages and vocabularies are only useful if they are used *for* something. In the next section, we will describe tools for using these languages and vocabularies, followed by some brief examples of successful use-cases that were built using these languages, tools, and vocabularies.

3.5.3 TOOLS

The above vocabularies, and others that are routinely being built these days in many different application areas, are far too large and complicated to be managed manually. The community has developed a large set of methods and tools for creating, managing, and deploying such large ontologies. Because of the rapidly evolving state of the art there is little point in putting any list of such tools in print, but they cover every aspect of an ontology's life cycle:

- *Creation* (either through knowledge acquisition and manual modelling, or through concept extraction from a corpus of text, or through machine learning from a large dataset)
- *Change management* (detecting which changes have occurred between versions, alerting for possible inconsistencies or redundancies this may have caused)
- *Modularization* (selecting the right subvocabulary for a given task)
- *Ontology alignment* (integration of multiple vocabularies for a single use)
- *Storage and querying* of very large data sets organized by an ontology (at the time of writing, ontology stores can handle in the order of billions of facts)
- *Reasoning* (drawing inferences from the given facts using the reasoning steps that are allowed under the formal semantics of the language)
- *Visualization* (visualizing large data sets organized by ontologies, either in the form of tree-diagrams or in the form of other diagrams such as cluster maps; see figure 3.1)

---

* http://www.ontologyportal.org/.
† http://wordnet.princeton.edu/.
‡ http://www.getty.edu/research/conducting_research/vocabularies/aat/.
§ http://www.iconclass.nl/.
¶ http://www.getty.edu/research/conducting_research/vocabularies/ulan/.

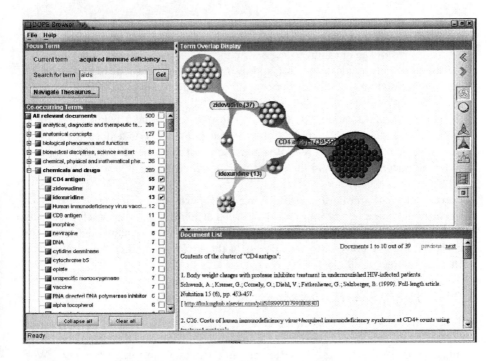

FIGURE 3.1 (See color insert.) Interface of the DOPE browser, showing ontology-based hierarchical and clustered display of search results using a single ontology across multiple collections (the color coding of individual objects reflects different types of publications. journal, conference, survey, etc.).

Many of these tools have outgrown the stage of academic prototyping and are available as commercial software, including support services.

3.5.4 ANNOTATION AND CLASSIFICATION TECHNIQUES

Of course, such large ontologies are only useful if used to describe and organize large data sets (the data sets between which we want to obtain semantic interoperability in the first place). This involves a task called *annotation* or *classification* in different parts of the community, but these are essentially the same task: given an item from a data set and given an ontology for the domain of the data set, decide which class(es) the data item must be assigned to. This task is of course crucial to the use of ontologies for solving semantic interoperability problems: if a data item is not assigned to a class in the ontology, it is not known to the receiving agent what to make of this data item, and which inferences to draw about this item.

This crucial task can be done either manually or automatically (in restricted domains).

- In a large number of domains, *manual* annotation will remain the dominant mode of classification for some time to come. In particular, for non-text items (sound, still images, video), automatic classification remains very hard (e.g., Snoek et al. 2007). In many audiovisual archives (e.g., in the cultural sector, satellite images, aerial photography, but also in medical applications), software uses an ontology to suggest annotations to users, but the final annotation is only made after user approval. The same holds for high-quality, long-term archives, such as national libraries (van Gendt et al. 2006).

• If an ontology is expressed in OWL DL, it is possible to determine the necessary and (sometimes) the sufficient properties that a data item must satisfy to belong to a certain class in the ontology. For text corpora, or for semistructured data sets, it is possible to analyze the properties of data items *automatically* and then determine to which class(es) they belong.

The tasks of annotating, classifying, and performing inference are often integrated into a single environment. Tools such as Sesame* or the Oracle† tool sets give a single integrated environment in which to write down semantic definitions of classes and their relations, store instances of those classes, and perform reasoning with such instances and classes.

3.5.5 Use-Cases, Scenarios, Applications

All the above languages, tools, and technology have been used to develop showcases in a variety of domains. Without any claim to completeness, we will sketch here a small number of use-cases that illustrate in particular the role of Semantic Web technology in semantic interoperability. Together, these use-cases do indeed make it credible that Semantic Web technology can indeed serve as the foundation for semantic interoperability in a worldwide information infrastructure.

eCulture browser over multiple art collections: In the eCulture browser‡ a number of art collections from major Dutch museums are manually annotated using a number of different ontologies (VRA, and the Getty thesauri AAT, ULAN, and TGN). Manual links that were established between these different ontologies then allowed a single-faceted browsing interface across the different collections. A similar result (although on a much smaller scale) was obtained in the STITCH browser,§ where illuminations from medieval manuscripts in the Dutch and French National Libraries are shown in a single interface, using automatically created links between the thesauri that were used to index the separate collections. An additional feature of the STITCH browser is that because one of the thesauri is multilingual, it becomes possible to search for manuscripts using search terms in a different language than the one with which they were annotated, using the multilingual thesaurus as a translation device. This is also very relevant in the spatial context; for example, in INSPIRE there should be support for 21 languages; see chapter 1.

---

* http://www.openrdf.org.
† http://www.oracle.com/technology/tech/semantic_technologies/index.html.
‡ http://e-culture.multimedian.nl/demo/search.
§ http://stitch.cs.vu.nl.

DOPE browser over multiple scientific literature collections: In the DOPE project (the Drug Ontology Project for Elsevier; Stuckenschmidt et al. 2004), a single large thesaurus (EMTREE, a commercial product by Elsevier, with 45,000 preferred terms and 190,000 synonyms to describe mainly drugs and diseases) was used to index a large body of scientific literature (5 million abstracts from the Medline database and another 500,000 full text papers from Elsevier's Science Direct collection). Annotation of papers and abstracts with EMTREE terms was done automatically using commercially available concept extract techniques. These semantically indexed heterogeneous collections could then be browsed using a single interface (see figure 3.1), which used the ontology for query disambiguation, hierarchical and clustered displays of search results, and query refinement.

The *Semantic Web Education and Outreach (SWEO) Interest Group* from W3C has published a collection of a few dozen use-cases of semantic technologies,* ranging from eGovernment to eCommerce, and from improved search to data integration, covering sectors as diverse as the automotive industry, the financial sector, the IT industry, the publishing world, and others.

Geosemantic applications: One of the use-cases in the SWEO collection describes the use of semantic technologies by the British Ordnance Survey† to cut the cost and improve the accuracy of data integration. By using an ontology of the Ordnance Survey's data, the semantic differences between different data sets are made explicit, thus facilitating data integration, both among OS data sets and between data sets of OS and its customers. An ontology has been built for the hydrology domain, as well as an administrative geography for Great Britain in RDF. An ontology for buildings and places is in progress.‡ Similar work on semantic translations of cadastral information has been reported in Hess and de Vries (2006). Closely related work, but aimed at the integration and chaining of geographic *services* as opposed to geographic data sets, is reported in Lemmens et al. (2006). Other work aiming at the integration of geographic services has earned a top ranking in the 2006 Semantic Web Challenge.§

3.6 THE MOST IMPORTANT CHALLENGE: ONTOLOGY MAPPING

Perhaps the most important challenge of all is how to deal with semantic interoperability across multiple ontologies; although a shared ontology gives a way to draw shared inferences that explicate the intended meaning of data items, this does require a *shared* ontology. Semantic interoperability across multiple ontologies requires these different ontologies to be aligned. This problem has been the subject of research in different fields over many decades. Different variants of the problem received names such as "record linkage" (dating back to Newcombe's work on linking patient

---

* http://www.w3.org/2001/sw/sweo/public/UseCases/.
† http://www.w3.org/2001/sw/sweo/public/UseCases/OrdSurvey/.
‡ All of these available at http://www.ordnancesurvey.co.uk/ontology.
§ http://www.laits.gmu.edu/geo/nga/.

records [Newcombe et al. 1959], and surveyed in Winkler [1999]), "schema integration" (Rahm and Bernstein 2001), and, more recently, "ontology mapping" (see the recent book by Euzenat and Shaivko 2007 for what is currently the best survey of the state of the art).

An important development in this historical progression is the move towards ever richer structure: the original record linkage problem was defined on simple strings that were names of record fields; the schema-integration problem already had the full relational model as input; while ontology mapping problems are defined on full hierarchical models plus rich axiomatizations. Each step in this progress has all the solutions of the previous steps at its disposal (as each later model subsumes the earlier ones), plus new methods that can exploit the richer structures of the objects to be aligned.

Current approaches to ontology mapping deploy a whole host of different methods, coming from very different areas. These can be categorized to distinguish linguistic, statistical, structural, and logical methods. The currently available best survey of ontology alignment techniques is by Euzenat and Shvaiko (2007).

Linguistic methods are directly rooted in the original record linkage work all the way back to the early 1960s. They try to exploit the linguistic labels attached to the concepts in source and target ontologies in order to discover potential matches. This can be as simple as basic stemming techniques or calculating Hamming distances, or can use specialized domain knowledge.

Statistical methods typically use *instance data* to determine correspondences between concepts: if there is a significant statistical correlation between the instances of a source-concept and a target-concept, there is reason to believe that these concepts are strongly related (by either a subsumption relation or perhaps even an equivalence relation). These approaches of course rely on the availability of a sufficiently large corpus of instances that are classified in both the source and the target ontology.

Structural methods exploit the graph structure of the source and target ontologies, and try to determine similarities between these structures, often in coordination with some of the other methods: if a source- and target-concept have similar linguistic labels, then dissimilarity of their graph neighborhoods can be used to detect homonym problems where purely linguistic methods would falsely declare a potential mapping.

Logical methods are perhaps most specific to mapping *ontologies* (instead of mapping record fields or database schemata). After all, in the time-honored phrase of Gruber (1993), ontologies are "*formal specifications* of a shared conceptualisation'" (my emphasis), and it makes sense to exploit this formalization of both source and target structures. A particularly interesting approach is to use a third ontology as background knowledge when mapping between a source and a target ontology: if relations can be established between source (resp. target) ontology and different parts of the background knowledge, then this induces a relation between source and target ontologies. A serious limitation to this approach is that many practical ontologies are rather at the semantically lightweight end of Uschold's spectrum (Uschold and Gruninger 1996), and thus do not carry much logical formalism with them.

Given the difficulty of the problem, and the amount of work already expended on it, it seems unlikely that the problem of ontology mapping will yield to a single solution. Instead, this seems more the kind of problem where many different partial solutions are needed. Currently, our toolbox of such partial solutions is already quite well stocked, and is still growing rapidly. However, a theory of which combination of partial solutions to apply in which circumstances is still lacking.

3.7 CONCLUSION

Undoubtedly, the problem of semantic integration is one of the key problems facing computer science today. Despite many years of work, this old problem is still open, and has actually acquired a new urgency now that other integration barriers (physical, syntactic) have been largely removed.

The ontology-based approach of the Semantic Web with its richer data models (logic-based hierarchical ontologies instead of the flat relational models), which allow rich inferences to be made, is a promising foundation for semantic interoperability in the information infrastructure.

REFERENCES

Bechhofer, S., F. van Harmelen, J. J. Hendler, I. Horrocks, D. McGuinness, P. Patel-Schneider, and S. Stein. 2004. OWL Web Ontology Language Reference, W3C Recommendation. February 10. http://www.w3.org/TR/owl-ref/.

Becket, D. 2004. RDF/XML Syntax Specification (Revised), W3C Recommendation. February 10. http://www.w3.org/TR/2004/REC-rdf-syntax-grammar-20040210/.

Cees, G., M. Snoek, B. Huurnink, L. Hollink, M. de Rijke, G. Schreiber, and M. Worring. 2007. Adding Semantics to Detectors for Video Retrieval. *IEEE Transactions on Multimedia* 9 (5):975–986.

Euzenat, J., and P. Shvaiko. 2007. *Ontology Matching*. Berlin: Springer-Verlag.

Gruber, T. 1993. A Translation Approach to Portable Ontologies. *Knowledge Acquisition* 5 (2):199–200.

Hayes, P. 2004. RDF Semantics, W3C Recommendation. February 10. http://www.w3.org/TR/rdf-mt/.

Hess, C., and M. de Vries. 2006. From Models to Data: A Prototype Query Translator for the Cadastral Domain. *Computers, Environment and Urban Systems* 30:529–542.

Lemmens, R., A. Wytzisk, R. de By, C. Granell, M. Gould, and P. van Oosterom. 2006. Integrating Semantic and Syntactic Descriptions to Chain Geographic Services. *IEEE Internet Computing* 10 (5):42–52.

Newcombe, H. B., J. M. Kennedy, S. J. Axford, and A. P. James. 1959. Automatic Linkage of Vital Records. *Science* 130:954–959.

Patel-Schneider, P., P. Hayes, and I. Horrocks. 2004. OWL Web Ontology Language, Semantics and Abstract Syntax, W3C Recommendation. February 10. http://www.w3.org/TR/owl-semantics/.

Rahm, E., and P. A. Bernstein. 2001. A Survey of Approaches to Automatic Schema Matching. *VLDB Journal: Very Large Data Bases* 10 (4):334–350.

Smith, B. 1982. Reflection and Semantics in a Procedural Language. Ph.D. thesis, Massachusetts Institute of Technology. Report MIT-LCS-TR-272.

Stuckenschmidt, H. , F. van Harmelen, A. de Waard, T. Scerri, R. Bhogal, J. van Buel, I. Crowlesmith, Ch. Fluit, A. Kampman, J. Broekstra, and E. van Mulligen. 2004. Exploring Large Document Repositories with RDF Technology: The DOPE Project. *IEEE Intelligent Systems* 19 (3):34–40.

Uschold, M., and M. Gruninger. 1996. Ontologies: Principles, Methods, and Applications. *Knowledge Engineering Review* 11 (2):93–155.

van Gendt, M., A. Isaac, L. van der Meij, and S. Schlobach. 2006. Semantic Web Techniques for Multiple Views on Heterogeneous Collections: A Case Study. *Proceedings of the 10th European Conference on Research and Advanced Technology for Digital Libraries* (ECDL 2006, Alicante, Spain, September 17–22, 2006), ed. Julio Gonzalo, Constantino Thanos, M. Felisa Verdejo, and Rafael C. Carrasco. Lecture Notes in Computer Science no. 4172, 426–437. Berlin: Springer Verlag.

Winkler, W. 1999. *The State of Record Linkage and Current Research Problems*. Technical report, Statistical Research Division, U.S. Bureau of the Census, Washington, DC.

4 Standardization and Modeling of Transportation Infrastructure Semantics
Experience from GDF, TransXML, OPAL, and Framework

Paul Scarponcini

CONTENTS

The move toward semantic systems necessitates a transition from human retrieval of information to machine understanding. This will be characterized by a move to persisting knowledge instead of information, represented by ontologies instead of database schemas, retrieved as individual chunks of information instead of entire documents, and processed by inference rules instead of operations. Guarino, for example, advocates a central role for ontologies to play in information systems, leading to his concept of "ontology-driven information systems" (Guarino 1998, 3).

So what is an ontology? Gruber defines it as a "specification of a conceptualization" (Gruber 1993). Guarino goes on to emphasize the need to distinguish between a language-dependent ontology and the language-independent conceptualization it characterizes.

In this chapter, an ontology is presented for each of several software standards and projects. Each is specified with a Unified Modeling Language (UML) model of concepts (classes), including their properties and relationships. Some might argue that UML class diagrams do not qualify as ontologies (because UML does not have formal semantics), though others may counter this by saying that there are different degrees of formality, as explained in Uschold and Gruninger (1996). Use of domain-specific data types, enumerations, and code lists to define domain values and the inclusion of definitions in the UML model (though not visible in the diagrams) or in an adjunct data dictionary adds to the formality for the presented standards and projects.

As each ontology is presented, the underlying conceptualization is explained. It is important to understand the conceptualization that each model supports in order to accept why each resultant ontology differs. This can also provide insight into how the ontologies can be integrated or at least made to be interoperable.

4.1   EARLY SPATIAL STANDARDS

Early spatial standards strived to achieve agreement on generic concepts of spatial data. These standards include the Open Geospatial Consortium (OGC) Simple Features Specification (OGC 1998), the ISO/IEC JTC 1/SC32 SQL/MM 13249-3 Spatial Standard (ISO/IEC 1999), and the first 20 parts of ISO TC211 (Geographic Information/Geomatics) (ISO 2002). These standards are purposefully independent of specific themes or domains, instead focusing on a generic feature construct. Structures to support feature properties and associations are provided without enumerating theme- or domain-specific properties or associations.

The ISO TC211 19100 series of geographic information standards "establishes a structured set of standards for information concerning objects or phenomena that are directly or indirectly associated with a location relative to the Earth" (ISO 2002, 7). This is accomplished using conceptual models containing abstractions of real-world features. Other standards, such as OGC Simple Features, SQL/MM Spatial, and GML (Geography Markup Language) (ISO 2005b), then specify how these conceptual models are implemented in a particular language, such as SQL or XML.

The first twenty 19100 standards focus on general geo-information concepts, which might apply to any domain. These include spatial and temporal schema, spatial referencing, cataloguing, metadata, quality, portrayal, and services. Of particular note is ISO 19109, Rules for Application Schema (ISO 2003). It specifies how domain-specific application schemas should be created in a consistent manner. Fundamental to this is the concept of a "feature." A feature is "an abstraction of real-world phenomena." A feature type then would be a description of a set of features that share the same attributes, operations, associations, constraints, and semantics. But ISO 19109 stops short of specifying any particular domain semantics; these are left up to the application schemas. In fact, it does not even specify specific feature types.

The TC211 standards use the Unified Modeling Language (UML) (Rumbaugh, Jacobson, and Booch 1999) class diagrams to document the conceptual models. Figure 4.1 is representative of this type of diagram, extracted from ISO 19109. Notice that most classes in this diagram have a stereotype (enclosed in guillemets) of Metaclass. This signifies that instances of these classes are themselves classes, not objects. Consequently, GF_FeatureType here will result in specific feature types in an application schema such as roads in a transportation application schema. Roads can

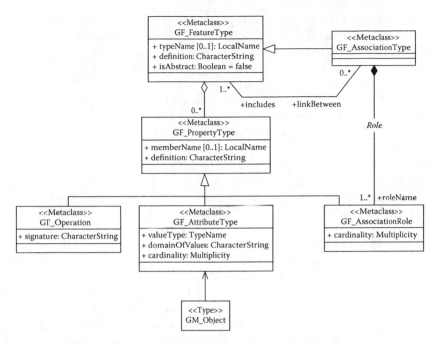

FIGURE 4.1   TC211 General Feature Model Extract from ISO 19109.

then have feature instances, such as Route 66, which would be an abstraction of the famous U.S. highway with that name.

A feature is an abstraction because it is impossible to include the real-world entity itself in a computer. Instead, only the aspects of the entity that are important to the application are included. These aspects become either properties of the feature or associations with other features. Properties can include attributes, operations, and association roles. Attributes can include, among other things, zero or more spatial representations of the feature, perhaps at different levels of precision. As with feature type, properties and associations are shown as metaclasses to allow each application schema to specify specific properties and associations particular to that application.

4.2   SEMANTIC TRANSPORTATION EFFORTS

The logical next step is to reach agreement within specific domains on the potential set of features, properties, and associations that are meaningful within that domain. Several such standardization efforts have resulted in a new set of application schemas in the transportation sector. They are presented here in breadth order, considering the number of domains and life cycle phases each addresses. All use UML to specify domain-specific semantics.

4.2.1   ISO 14825: Geographic Data Files

The Geographic Data Files (GDF 4.0, ISO IS 14825:1996) have been developed by ISO TC204 (Intelligent Transportation Systems) (ISO 1996). The GDF focuses on the operational phase within the roadway domain. It contains feature, attribute, and relationship catalogues that specify the types of features needed to describe map databases for in-car navigation systems and their properties and associations, respectively. The conceptualization here is focused on roads, but only those aspects of roads that are pertinent to car navigation.

Consistent with being an ISO 19109 application schema, the GDF defines feature types. Figure 4.2, from the working draft of GDF 5.0 (Scarponcini and Hiestermann 2006), shows the first level in the GDF feature classification scheme. The full taxonomy, including some 170 instantiable feature types, is the GDF Feature Catalog. Each feature type instance of ISO 19109 GF_FeatureType is so designated with Feature as its stereotype.

The GDF Attribute Catalog specifies attributes for each feature type. As in the example in figure 4.3, which shows the GDF feature "Structure," each attribute has a name, multiplicity, and type, as in "maxHeightAllowed[0..1]: MaxHeightMeasure." By TC211 convention, attribute names are always in lowerCamelCase. Multiplicities tell how many values for this attribute are allowed per feature instance. A multiplicity of [0..1] following the attribute name means zero to one, that is, the attribute is optional. A multiplicity of [0..*] signifies that the attribute is optional and that any number of values would be appropriate. A multiplicity of [1..*] signifies that the attribute is mandatory and that any number of values (at least one) would be appropriate. Lack of an explicit multiplicity value is equivalent to [1], that is, the attribute is mandatory and exactly one value is required for each instance of the feature type.

The attribute type, in UpperCamelCase, may be a simple data type like CharacterString, Number, DateTime, or Boolean. Alternatively, it may be a type specific to

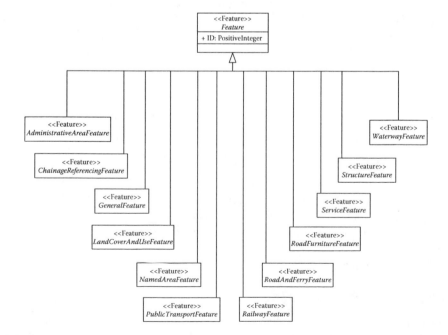

FIGURE 4.2   GDF feature type hierarchy.

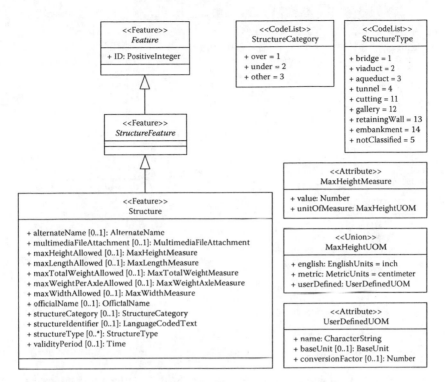

FIGURE 4.3   GDF feature "Structure" with attributes.

GDF. These types are specified with stereotypes of either Enumeration, CodeList, DataType, Union, or Attribute. All but Attribute are adopted from TC211. An enumeration is a domain of allowable values for an attribute, provided as a fixed list of alternative values. The values are sometimes accompanied by a shorthand code value, shown as a default value following an equals sign (=) in accordance with UML syntax for initial value. A code list is similar to an enumeration, but allows for expanding the list of valid values. Data type is a nonsimple type usually defined as a combination of other attributes. It does not have a unique identity of its own. Union is a type consisting of several alternatives (listed as member attributes) representing a discriminating union of these alternatives, that is, only one may be selected. The Attribute stereotype is used instead of DataType in GDF to signify that the value does have an identity if it is an attribute of a feature or relationship and not an attribute of another attribute or part of a composite attribute.

The Attribute Catalog is perhaps the most significant semantic aspect of GDF. It tells what information is significant about each feature. Moreover, it provides meaning to this information by specifying the allowable values or value types. The revised GDF Attribute Catalog in GDF 5.0 has over 450 attributes with 70 enumerations and code lists.

The GDF Relationship Catalog specifies over 50 relationship types that would qualify as instances of the ISO 19109 GF_AssociationType metaclass. Figure 4.4 diagrams one such association between features, the "Fork" relationship. In GDF, relationship classes are distinguishable by their Relationship stereotype. Attributes

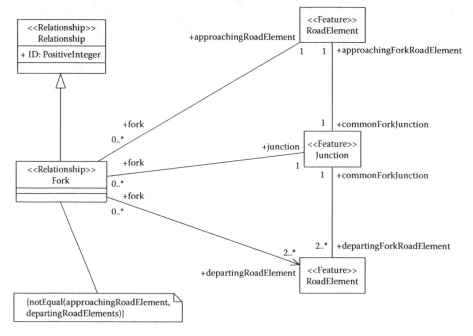

FIGURE 4.4   GDF relationship "Fork" with roles.

of the relationship are shown in the middle part of the relationship class rectangle. The feature types involved in a relationship are shown connected with UML association lines. These lines are adorned with multiplicities and roles at either end.

Multiplicities have similar values as those used with attributes. Here they specify how many instances of the type nearest the multiplicity in the figure can exist in each association with the type at the opposite side of the association line. For example, in the Figure 4.4, each fork must have exactly one road element playing the role of approaching road element and any number of (at least two) departing road elements. The role label (e.g., "approachingRoadElement") is an instance of the ISO 19109 GF_Association Role in figure 4.1. That is, it is a property of the feature type fork relationship (from figure 4.1, GF_AssociationType is a subtype of GF_FeatureType, so GDF relationship types are also TC211 feature types). Properties in the form of relationship roles add to the semantic richness of GDF.

In GDF 4.0, the conceptual model was presented using NIAM (Nijssens Information Analysis Method) (ISO 1996). A significant effort is being made in GDF 5.0 to recast the conceptual model using UML class diagrams, consistent with TC211 conventions. These diagrams have proven to be easier to understand. For example, NIAM uses a combination of dots and underlining to specify the cardinality of associations, whereas UML uses numbers like "0..1" to mean zero or one. UML also provides more detailed semantic content than NIAM, with the addition of roles and attribute domain specification (data types, enumerations, and code lists).

Because all of the diagrams are generated from a single UML model, it is now possible to achieve consistency between the diagrams, which was not possible before when the individual NIAM diagrams were drawn independently with Microsoft

Word graphics. For example, when attempting to add attributes to the airport feature type, it became apparent that in GDF 4.0 there were two feature types with the same name. Airport existed as both a type of land cover as well as a type of service. The former has now been changed to airport area to represent the land area containing the airport service.

The UML conceptual model also allows multiple future applications, including those beyond car navigation, to be based on a common roadway information model. The model can be used for the implementation of information systems according to a model-driven architecture approach. The same model can be the basis of multiple realizations beyond the current, single, native file format. In fact, GDF 5.0 will also include a database schema (SQL Data Definition Language) and an exchange format (XML schema and, if time permits, GML3). Though the UML provides the basis for these implementations and helps ensure their consistency, the SQL DDL and XML schema were manually derived from the UML class diagrams.

4.2.2 NCHRP 20-64 Project: TransXML

The NCHRP 20-64 Project, TransXML, took a broader view of transportation systems (Ziering, Harrison, and Scarponcini 2007). NCHRP is the National Cooperative Highway Research Program under the U.S. Transportation Research Board, which administers research projects. TransXML looked across the various life cycle phases of a transportation facility, including planning, design, construction, operations, and maintenance. TransXML was quite successful in helping to bridge the semantic gap between design and construction by standardizing contract pay items as they evolve from design through bidding and into construction.

A fundamental concept during the design and construction phases is the pay item, the individual component that road construction and improvement projects are broken down into (so named because they dictate how the contractor will be paid for the work performed). An example of a pay item is sidewalk. If there is sidewalk to be constructed as part of the project, then the quantity of this pay item is estimated during design and measured after construction. The contractor is then paid based on the agreed-upon unit price for sidewalk. Each pay item has a description (4.5 foot wide concrete sidewalk, four inches in thickness). Its unit property (linear feet in this example) specifies how it will be measured for payment.

The TransXML team decided to develop a UML conceptual model prior to developing any XML code. The model was divided into packages based on life cycle phase based tasks. The packages that used pay items included Design Project, Bid Package, and Construction Progress. As they investigated the semantics of what a pay item was, they discovered that it was something different in each phase. During design, the pay item represented part of the as-designed project and had a quantity and unit price estimated by the designer. Once the project was put out to bid, prospective contractors would submit the unit price they were willing to charge for each pay item, based on the quantities they perceived would be needed to complete the job. The winning contractor's unit prices become the agreed-upon amount they would actually be paid. During construction, progress is measured by determining the installed quantity of each pay item. The contractor may be given partial payment based on the actual quantity installed and the contract unit prices. The final project cost would be the sum of the completed quantities times the contract unit price for each. So there were really three different pay items, as shown in figure 4.5: design project pay item, contract pay item, and construction project pay item. These are, of course, all related by associations. Additionally, each has associations to other features relevant during each particular life cycle phase.

On further investigation, it became apparent that a fourth pay item type was needed. Most transportation departments keep a list of standard pay items they expect to have on their projects. The designer selects from this list of reference pay items when creating the design project pay items. This ensures consistency across projects, such as establishing how each is measured for payment. It also allows estimators to keep records of contractors' bid unit prices across projects in order to more effectively estimate the expected bid prices on subsequent projects.

Once again, UML models helped illuminate the semantic differences between reference, design project, contract, and construction project pay items, based on their respective attributes and associations. A pay item is conceptualized differently by different people and at different times during the project. These semantic distinctions proved to be essential in clearly articulating the transition from design to construction.

4.2.3 PROJECT OPAL

The next step is to look beyond a single domain or information community. Within Bentley Systems, two new products are available through acquisition. The Location Data Manager (LDM) focuses on data storage for roadway information, whereas OPTRAM excels at presentation for rail system information. In order to integrate

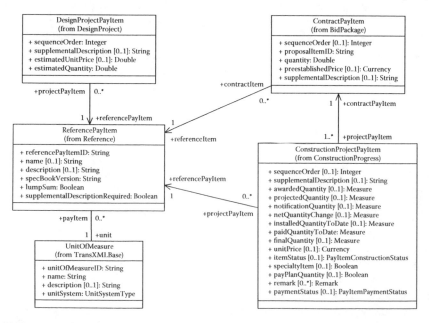

FIGURE 4.5   TransXML pay item evolution.

the two products, Bentley first developed independent, semantically rich UML models of the existing concepts employed in each system, complete with definitions and attribute domain specifications. From these, an overall conceptual model was developed, consolidating the two models into one (Scarponcini 2007). Agreeing on combined semantics has resulted in a vision for bringing the two products together, known internally as OPAL.

When integrating the conceptual models, several options were available. If both models contained the same concept that differed only in concept name, a mutually agreed upon name was substituted. If the concept in one model subsumed a concept in the other model, it was selected and the additional attribution was investigated further to determine if it should be made optional. If the concepts were similar but differed in some attribution or associations, a higher-level concept was introduced and the existing concepts were subtyped off of it. If two concepts were mutually exclusive, both were retained.

Figure 4.6 demonstrates these options in the synthesis of LDM and OPTRAM concepts into OPAL. The concept of a linear element evolved from the Bentley Generalized Model for Linear Referencing (Scarponcini 2002) and has been standardized in the TC211 standard for Linear Reference Systems in ISO 19133 (ISO 2005a). A linear element is any one-dimensional object that can be measured, that is, it supports the ILinearElement interface. This interface includes operations that allow translation between linearly referenced locations having spatially equal locations but which are specified using different linear referencing methods or as being along different linear elements. Both linear element and the ILinearElement interface were adopted by OPAL as concepts applicable to both road and rail. The road-based

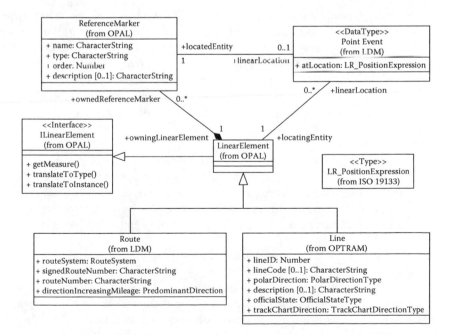

FIGURE 4.6   OPAL reference marker.

route feature type as it currently exists in LDM and the rail-based line feature from OPTRAM are adopted into OPAL as subtypes of linear elements because of their common linear behavior.

Both LDM and OPTRAM had versions of reference markers, as both road and rail industries use them to physically mark locations along the road or rail line, respectively. But each system defined them with different properties and associations, that is, with different conceptualizations. So a more generalized concept of reference maker was created for OPAL that could be easily adopted by either product. LDM had a concept of a point event as a method for specifying the location of something along a linear element. This works for reference markers, as their location must be known before they can be used to define the location of something else. For the point event, the reference marker plays the role of a located entity and the route or line can play the role of a locating entity because they are both subtypes of linear elements. The actual location of the reference marker along the line or route is given by the point event's "at location" attribute, which uses the ISO 19133 standard to specify its linearly located position.

UML proved to be helpful in first capturing the semantics of each of the two original products. This enabled the two development organizations to understand the other product more easily. In fact, it helped them understand their own product as well, because many of the developers only worked on a small area within each product. UML then proved even more useful in proposing an overall conceptual model to support the vision for product migration.

4.2.4 FRAMEWORK DATA CONTENT STANDARD

The U.S. Federal Geographic Data Committee (FGDC) has gone one step further by creating a standard that spans multiple domains, of which transportation is only one (FGDC 2006). Developed with the goal of supporting geographic data transfer between government agencies, the Framework Data Content Standard covers seven themes, including cadastral, orthoimagery, elevation, geodetic control, governmental unit and other geographic area boundaries, hydrography, and transportation. As the seven domain-specific teams worked on their individual parts, a harmonization team was convened (Kerestesy 2003) whose tasks were:

1. To harmonize the models, normative text, normative references, definitions, and data dictionaries of separate working drafts into an NSDI Framework Standard composed of a base part and subparts representing theme content.
2. To ensure consistency of the resulting standard with existing international, national, and federal standards, and with standards under development.

For the base part of the standard, which was responsible for defining the concept of a feature, the TC211 definition was adopted.

The transportation theme has five modal parts: road, rail, transit, air, and navigable waterways. Harmonizing the five parts into a consistent approach for transportation proved to be a significant but successful undertaking, resulting in a harmonized transportation base model.

The road part was done first. The Road Modeling Advisory Team (MAT) began by looking at existing standards and models. Each appeared to be too specific in terms of the applications it was attempting to support, whereas the MAT was looking for a more widely applicable generalized model. After unsuccessfully attempting to merge the individual models, they decided to start with a clean sheet of paper. It is true that the existing individual models will have to be mapped to this new model, but this should be attainable if the resultant model is indeed a more generalized conceptualization.

The biggest struggle was to agree on the semantics of what a road is. Where does one road start and the next one begin? Is it a single road with dual carriageways or separate, directed roads? Or is each lane significant?

Unlike discrete entities found in other domains, like a person in a human resources domain, a road is a continuous element. Its attributes can vary in value as you travel along the road. Each person thinks of a road as having a set of attributes pertinent to his or her particular application. Consequently, each application has a different concept of what a road is and where it begins and ends.

To deal with this apparent semantic incompatibility, the MAT decided to have a road segment (RoadSeg) as an arbitrary length of roadway. RoadSegs can be bounded by RoadPoints. The end user can decide where to locate RoadPoints and therefore how to segment roadways into RoadSegs. RoadSeg is therefore a higher-level concept including just those aspects of a road that span all of the more detailed conceptualizations at a lower level: they represent a length of physical roadway without being constrained by any particular set of physical roadway characteristics.

To support the lower-level conceptualizations while still maintaining a single set of RoadSegs, linear events were introduced. Linear events along the RoadSeg accommodate attributes with values, which potentially vary along the individual RoadSeg. For example, the speed limit might change 40% along the RoadSeg. Rather than forcing the RoadSeg to be split into two where the speed limit change occurs, one merely creates two linear events, one for each speed limit value, indicating where along the single RoadSeg each event applies. The alternative would have been to segment the road network wherever the value of any attribute changes. This approach would be inefficient, because you would have to repeat all the values that did not change at this location. Furthermore, it is impossible to predict a priori all of the possible attributes that might be part of anyone's conceptualization.

The physical road can be used for multiple purposes. For example, a RoadSeg may be part of an administrative route, like U.S. Route 66. The same RoadSeg can be part of bus route 27. Regardless of the number of times the RoadSeg is used, it still has the same two speed limits. So RoadPaths were introduced to model the use of one or more, whole or partial RoadSegs, inspired by the database design principle of normalization. It is then possible to determine the speed limit along Route 66 or bus route 27 from the underlying attribute defined along the same RoadSeg used by the two coincident routes. Figure 4.7 shows the properties and associations for the RoadSeg, RoadPoint, and RoadPath concepts.

This same conceptualization was then proposed to the Rail MAT. They decided to adopt similar concepts: RailSeg, RailPoint, and RailPath. They were even able to accommodate both the marketing view of RailSegs bounded by stations and the

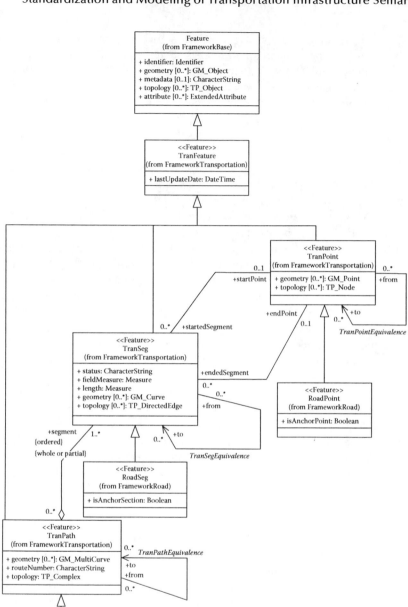

FIGURE 4.7   Framework transportation base and road.

engineering view of RailSegs bounded by points of switch, that is, individual track segments.

The Transportation Base team was then assembled to harmonize the various modal parts. They abstracted the Road and Rail MAT concepts into the TranSeg,

TranPoint, and TranPath types shown in figure 4.7 but labeled as coming from FrameworkTransportation. They were even able to accommodate multimodal transportation by allowing TranPoints and TranPaths to be instantiable. A TranPoint instance can then represent an intermodal node, connecting mode-specific Segs. A multimodal route can be defined as a TranPath along mode-specific Segs.

The transit mode also adapted the harmonized transportation segmentation model. Navigable water tried, but it proved to be too comprehensive for their simpler needs. The conceptualization did not apply to the air mode, which focused on (non-linear) airports.

A single comprehensive UML model was developed for the entire standard, from which part- and theme-specific diagrams were generated and published. This helped ensure consistency across the various parts and themes. Figure 4.7 was extracted from this model, showing classes contributed by three parts: framework base, transportation base, and road.

The U.S. Department of Homeland Security (DHS) has subsequently chosen to include the Framework standard themes in the first release of their Geospatial Data Model (DHS 2006). They have augmented these with additional domains.

4.3 FUTURE OPPORTUNITY

UML has played a significant role in capturing and documenting semantics in all of these projects. The projects include standardizing a single domain for a single application (GDF), a single domain across multiple life cycle phases (TransXML), two related domains (OPAL), and multiple disparate domains (Framework). In each case, the individual conceptualization drove the resultant ontology. Moving from the narrowly focused GDF conceptualization of car roads supporting car navigation to the more general Framework conceptualization aimed at supporting the transfer of spatial data between government agencies results in increasing ontology levels resulting from the increasingly generalized conceptualizations.

Future standards may rely on more formalized representation schemes than UML, such as OWL. ISO TC211 has issued a New Work Item Proposal, the scope indicated as "preliminary work to collect and compile information, and to investigate how ontology and semantic web approaches can benefit ISO/TC 211 objectives" (ISO 2007, 1). The project also intends to investigate the translation of some UML models into OWL and other structures for ontology. The first meeting was held in Rome in June, 2007.

REFERENCES

DHS. 2006. *DHS Geospatial Data Model*, version 1.1. Washington, DC: U.S. Department of Homeland Security.
FGDC. 2006. *Information Technology—Geographic Information, Framework Data Content Standard, Parts 0-7*. Federal Geographic Data Committee (as submitted to American National Standard Institute), Washington, DC.
Gruber, T. R. 1993. A Translation Approach to Portable Ontology Specifications. *Knowledge Acquisition* 5:199–220.

Guarino, N. 1998. Formal Ontology and Information Systems. In *Formal Ontology and Information Systems (FOIS '98)*, ed. Nicola Guarino. Amsterdam: IOS Press.

ISO. 1996. *ISO IS 14825:1996, Intelligent Transport Systems—Geographic Data Files—Overall Data Specification.* Geneva: International Organization for Standardization.

ISO. 2002. *ISO DIS 19101:2002, Geographic Information—Reference Model.* Geneva: International Organization for Standardization.

ISO. 2003. *ISO IS 19109:2003, Geographic Information—Rules for Application Schema.* Geneva: International Organization for Standardization.

ISO. 2005a. *ISO IS 19133:2005, Geographic Information—Location Based Services—Tracking and Navigation.* Clause 6.6 Linear Reference Systems. Geneva: International Organization for Standardization.

ISO. 2005b. *ISO DIS 19136:2005, Geographic Information—Geography Markup Language.* Geneva: International Organization for Standardization.

ISO. 2007. *ISO New Work Item Proposal, Geographic Information—Ontology.* ISO document 211n2163. Geneva: International Organization for Standardization.

ISO/IEC. 1999. *ISO/IEC IS 13249-3:1999(E) Information Technology—Database Languages—SQL Multimedia and Application Packages—Part 3: Spatial.* Geneva: International Organization for Standardization.

OGC. 1998. *OpenGIS® Simple Features Specification for SQL Revision 1.0.* Wayland, MA: Open GIS Consortium, Inc.

Kerestesy, L. 2003. *Harmonization Team Charter.* Geospatial One-Stop NSDI Framework Standards' Development project document. Washington, DC: U.S. Federal Geographic Data Committee.

Rumbaugh, J., I. Jacobson, and G. Booch. 1999. *The Unified Modeling Language Reference Manual.* Reading, MA: Addison-Wesley.

Scarponcini, P. 2002. Generalized Model for Linear Referencing in Transportation. *Geoinformatica* 6 (1):35–55.

Scarponcini, P. 2007. *OPAL Unified Model Approach: Consolidated Conceptual Framework.* Littleton, CO: Bentley Systems, Inc.

Scarponcini, P., and V. Hiestermann. 2006. *XGDF UML Base Document,* version 1.5. Working Draft document from ISO TC204 Task Group 3.1.3.

Uschold, M., and M. Gruninger. 1996. Ontologies: Principles, Methods, and Applications. *Knowledge Engineering Review* 2 (11):2.4.

Ziering, E., F. Harrison, and P. Scarponcini. 2007. *TransXML: XML Schemas for Exchange of Transportation Data.* NCHRP Report 576. Washington, DC: Transportation Research Board.

5 How Earth Science Can Contribute to and Benefit from the Spatial Information Infrastructure

Andrew Woolf and Stefano Nativi

CONTENTS

The emergence of spatial information infrastructures finds resonance with related developments in earth science. These include the evolution of informatics as an important subdiscipline and the application of advanced grid technology in service-oriented, information-rich earth science applications.

The growing area of environmental informatics is concerned with providing integrated access to a range of advanced information and processing resources for the environment. The Biosphere Data Project at the University of California Berkeley has defined environmental informatics as: "an emerging field centring around the development of standards and protocols, both technical and institutional, for sharing and integrating environmental data and information" (Biosphere Web site). A similar definition is provided by the U.K. Natural Environment Research Council (NERC science topics, Topic 16): "Research and system development focusing on the environmental sciences relating to the creation, collection, storage, processing, modelling, interpretation, display and dissemination of data and information." Both definitions illustrate a significant overlap with those systems more broadly described as spatial information infrastructures (SII).

The U.S. and European premier earth science unions are recognizing the scientific importance of such infrastructures—the American Geophysical Union has established an Earth and Space Sciences Informatics Focus group, and the European Geosciences Union is establishing a new scientific division on Earth and Space Science Informatics. Moreover, a survey of abstracts at meetings of these bodies (see, for instance, figure 5.1) indicates a significant increase in recent years of research papers related to ISO and OGC standards, and associated metadata, data and service models—all essential elements of the SII.

Section 5.1 considers the synergy that exists between earth science informatics and SII, and section 5.2 examines the special case of advanced earth science grid infrastructures. Sections 5.3 to 5.6 consider some of the particular challenges of earth science for SII, including information modelling, temporal aspects, "coverage" data and observations. Benefits flowing to the earth sciences from SII are considered in section 5.7, with some examples provided in section 5.8. Finally, conclusions are outlined in section 5.9.

5.1   A SYNERGY WITH SPATIAL INFORMATION INFRASTRUCTURES

Examining the scientific requirements of advanced applications in environmental informatics provides a clear rationale for the synergy that exists with SII. The most pressing environmental problems demand an integrated modelling of coupled physical processes, global data sets and a multidisciplinary coordinated approach (e.g. biologists working together with climatologists to determine the impact of warming on species distributions). These three characteristics map well onto properties of SIIs. The coupling between physical processes required in environmental simulations requires an information infrastructure where data from different geospheres (atmosphere, ocean, cryosphere, biosphere) may be integrated into common analysis

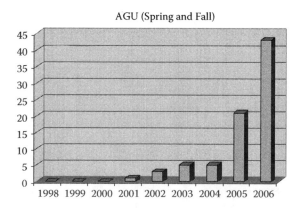

FIGURE 5.1   AGU presentation abstracts mentioning SII-related keywords (OGC, WFS, WCS, WMS, ISO, GML).

tools. The SII approach defines a canonical information modelling approach with mappings and translations from legacy formats onto common structures. The global nature of earth science matches well the approach of SII to federating resources into large-scale or global infrastructures (e.g. GEOSS). Finally, the multithematic approach of SII (e.g. INSPIRE) is entirely consistent with the increasingly important requirement of earth science to adopt a multidisciplinary approach—problems of biodiversity and climate change require information from many fields to be integrated together.

One of the key drivers behind the development of SIIs such as the European INSPIRE initiative is precisely in order to facilitate a political and legislative response to environmental problems. These include issues of air and water quality, climate change, energy use, etc. Such policymaking requires access to earth science data and information as primary material, both as evidence and for policy guidance. Thus, SII may be seen as a necessary outcome of the attempt to solve the most pressing environmental earth science problems.

5.2   GRID INFRASTRUCTURES FOR EARTH SCIENCE

Advanced e-infrastructures (or cyber-infrastructures) are supporting the formation and operation of an earth system science community, based on multidisciplinary knowledge integration. The science gateway program of the U.S. TeraGrid project (TeraGrid site) and the actions funded to accompany the EU EGEE (Enabling Grids for E-sciencE) project (EGEE site), are examples of the impact of this trend on current e-infrastructures.

These developments require scaling from specific and monolithic systems (datacentric) towards independent and modular (service-oriented) information systems. In fact, such an infrastructure must provide scientists, researchers and decision makers with a persistent set of independent services and information that scientists can integrate into a range of more complex analyses (Foster and Kesselman 2006). The recent revolution in information and communication technologies, such as MDA (Model-Driven Architecture), SOA (Service-Oriented Architecture), semistructured data models and encodings, etc. and consequent infrastructures (e.g. Internet, GRID, etc.), now provides the basis for making significant steps towards these platforms.

An earth science system-level approach is the key component of several international initiatives and programs dealing with environmental monitoring, risk management and security. Two of the most important initiatives are GEOSS (Global Earth Observation System of Systems) and GMES (Global Monitoring for Environment and Security).

GMES was endorsed by the EU Commission in 2001 and will create value-added services that support decision makers in crisis prevention and mitigation, and in environmental and security management. Civil protection (CP) is one of the most important service categories. Many CP and GMES applications require the integration of infrastructures involving many actors (civil protection systems, public authorities, research agencies, etc.) and coordinated sharing of information and services.

Grid technologies support the sharing and coordinated use of diverse resources in dynamic virtual organizations. These resources may be computational, data or even instruments and large-scale facilities. This implies the creation, from geographically and organizationally distributed components, of virtual computing systems that are sufficiently integrated to deliver the desired quality of service (QoS) (Foster and Kesselman 1999; Foster et al. 2001). Advanced security models (for authentication, authorization and accounting) provide access control and implement resource sharing policies. Grid concepts are critically important for computing not primarily as a means of enhancing capability, but rather as a solution to new challenges relating to the construction of reliable, scalable and secure distributed systems (Foster et al. 2007). Many earth science application domains increasingly rely on large-scale, complex and expensive data and computing resources, and progress often requires collaborative working. Consequently, the adoption of a grid-based infrastructure seems a natural choice to start building a cooperative platform for supporting GMES and CP applications.

However, GMES applications have specific and advanced requirements, such as (CYCLOPS 2006): rapid (real-time) access to information and models (especially during emergency situations), marshalling and control of sensor networks and processing chains, sharing of large-volume dynamic spatial data (e.g. remotely sensed satellite observations), secure interaction with military resources and distributed image processing from acquisition through to decision support. Moreover, the GMES community is focused on strategic applications and high-level concepts (e.g. storm surge models, seabed classification, seismic risk areas, data mining services, data fusion, etc.).

On the other hand, the grid community provides raw technological capability provisions (i.e. storage, computational power, networking, etc.). Thus, grid infrastructures need enhancement to fully support CP and GMES applications. Advanced spatial information services can play an important role in addressing the semantic mismatch, providing the functionalities required by GMES applications. Figure 5.2 depicts this interoperability framework, as envisioned and analyzed by CYCLOPS: grid infrastructures can provide the resources (e.g. computing, data, network, sensor, etc.) required by the GMES/CP virtual organization. On the top of this infrastructure

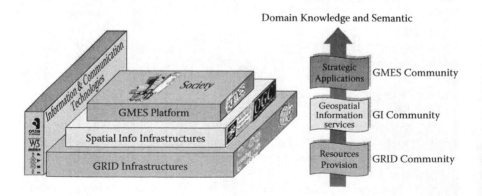

FIGURE 5.2 (See color insert.) The CYCLOPS vision for a possible GMES/CP interoperability framework.

a SII can provide high-level spatial information services that underpin the GMES/ CP community strategic applications.

We consider below two examples of grid infrastructures applied in the earth sciences, CYCLOPS and NERC DataGrid.

5.2.1 CYCLOPS, A GRID APPLICATION FOR GMES

The European Union (EU) Cyber-Infrastructure for Civil Protection Operative Procedures (CYCLOPS) project (CYCLOPS site) aims to interconnect the GMES and Grid communities, investigating the development of a specific European Civil Protection e-infrastructure. Presently, the EGEE Grid consists of over 30,000 CPUs available to users 24 hours a day, 7 days a week, in addition to about 5 Petabytes of storage, and maintains 30,000 concurrent jobs on average (EGEE site).

An essential objective for CYCLOPS is to cross-disseminate the approaches, requirements and visions of the two communities to develop an advanced information infrastructure as represented by figure 5.3. This infrastructure is conceived to serve the CP and GMES strategic sectors (CYCLOPS 2006).

The architecture of figure 5.3 shows an infrastructure based on existing EGEE middleware providing basic services for the coordinated sharing of processing and data system resources. Environmental monitoring resources like sensors need to be *grid-enabled* and "virtualized" through specific services (e.g. sensor discovery, description, access for acquisition and control, etc.).

On top of the grid middleware, specific persistent earth science grid services must be implemented to build a suitable GMES/CP application platform; these are of two main types:

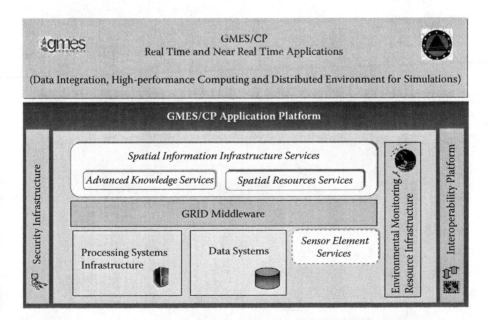

FIGURE 5.3   (See color insert.) The CYCLOPS application platform for GMES/CP.

- Advanced knowledge services (e.g. knowledge extraction and management services; decision support services; mediation services for multicultural, multilingual and multidisciplinary contents; QoS management; orchestration services; etc.)
- Geospatial* information infrastructure services for spatial resource management (e.g. value-added processing, discovery and cataloguing, accessing, configuration and control, etc.)

An advanced security infrastructure provides the security and policy services required for handling the complex data policies typical of dual systems (civil/military). They must satisfy the strict requirements of integrity, confidentiality and data/services access control.

To complete their tasks, many GMES/CP applications should be able to interact with external infrastructures, such as security systems and e-government infrastructures. Thus, an interoperability infrastructure completes the platform.

5.2.2 NERC DataGrid

NERC DataGrid (NDG site) is a U.K. project aimed at integrating access to a wide range of environmental data within the United Kingdom (Lawrence et al. 2004; Woolf et al. 2004). It is focused initially on atmospheric and oceanographic data—both model simulation and real-world observation, including remote-sensed imagery; see figure 5.4. The project has committed to a standards-based approach conformant to SII architectures, and has developed an information model, data services, and a novel security mechanism. The project's technology is being exploited by other

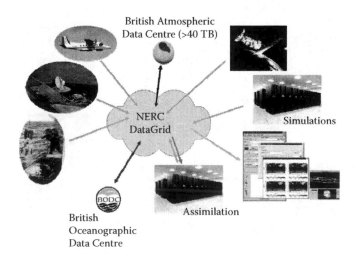

FIGURE 5.4   (See color insert.) NERC DataGrid.

* *Geospatial* is a term widely used to describe the combination of spatial software and analytical methods with terrestrial or geographic datasets (Wikipedia).

projects, including the high-profile Data Distribution Centre (DDC site) of the Intergovernmental Panel on Climate Change (IPCC).

A number of issues have been identified and are being actively pursued by the NDG in applying SII standards to earth science data infrastructures (Woolf et al. 2005; Lowe et al. 2006). First, the use of earth science vertical coordinates is different than the metric length-based coordinates supported through ISO 19111. For instance, vertical coordinates may be based on pressure or density, or terrain-following surfaces. GML's implementation of ISO 19123 coverage structures is limited. Various registers are required for scalability in SII (e.g. feature catalogues, units of measure, coordinate reference systems, controlled vocabularies, etc.). Implementations of the Open Geospatial Consortium's Web Map Service (WMS) and Web Coverage Service (WCS) need to support slicing in directions other than horizontal (e.g. vertical slices, or Hovmüller slices in time, or even temporal animation). Finally, best practice needs to be determined for referencing of binary file-based content from GML.

5.3 INFORMATION MODELLING: APPROACHES IN SII AND EARTH SCIENCE

The information modelling approach used in SIIs is novel—for the most part—to the earth science community, and there are significant challenges in bridging the conceptual gap.

The model-driven approach within SII is now formalized in a raft of standards from ISO TC211 (Woolf et al. 2005), and has been applied and proven in a number of domains (e.g. cadastral, administrative boundaries, etc.). The procedure is described in ISO 19101 and ISO 19109, and is illustrated in figure 5.5. First, a formal model is developed in UML (according to the UML profile defined in ISO 19103), describing the logical structure and semantic content of a data set. This model is the *Application*

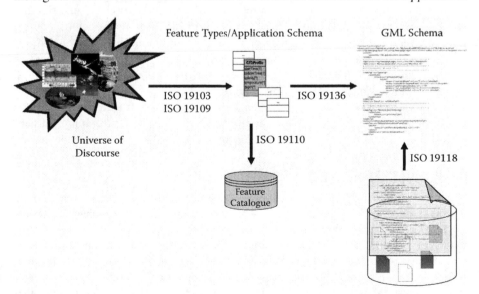

FIGURE 5.5 (See color insert.) Information modelling approach of ISO TC211.

Schema and uses *feature types* (ISO 19109) for the important information classes in a domain. These feature types can be registered into a *Feature Catalogue* for reuse (ISO 19110). The UML Application Schema can be transformed into a *GML Application Schema* using the procedure outlined in ISO 19136. A data set can then be *encoded* as a GML instance document according to the GML schema (ISO 19118). For imagery and gridded data, both ISO 19121 and ISO 19129 have envisaged an encoding model that combines a canonical (GML) representation with file-based data (generally a binary encoding) for efficiency.

In earth science there have been just a few applications of this full-cycle modelling approach, notably in the geosciences (GeoSciML; GeoSciML site) and atmosphere/ ocean (CSML; Woolf et al. 2006). The traditional approach to information management in many earth science disciplines has been *file-based* rather than *content-based*, that is, the focus is on the file format, rather than on the logical semantic structure of the data itself.

5.4   TEMPORAL ASPECTS: AN EARTH SCIENCE CHALLENGE FOR SII

Geographic information can be defined as information concerning phenomena implicitly or explicitly associated with a location relative to the Earth (ISO 19101). It is straightforward to recognize two significant themes: observed phenomena and earth locations. Due to the intrinsic nature of earth science and the associated acquisition technologies (e.g. multiparametric remote sensing techniques), earth science data sets focus on information related to complex phenomena, with earth location aspects traditionally kept as simple as possible (Nativi et al. 2004).

Conversely, time is essential for understanding earth science phenomena. It can be expressed in units ranging from seconds (e.g. rainfall variations measured by a sequence of radar scans) to centuries (climatological variations calculated through complex models). Both running clock (e.g. experiment time) and epoch-based (e.g. date and time and ordinal systems) approaches are commonly used. For earth science data, time location and evolution of observed phenomena are as important as spatial location (Nativi et al. 2004). Neglecting the temporal aspects could constitute an unacceptable simplification (Nativi and Ross 2007).

Therefore, SII must be able to manage time (e.g. temporal extent metadata and temporal contextualization of information), especially time series (e.g. plume trajectories), and must also deal with the problem of data related to previous system states.

5.4.1   TEMPORAL EXTENT IN DISCOVERY METADATA

Temporal-extent metadata to discover geospatial resources is recognized to be essential for several important information communities (e.g. atmospheric sciences, oceanography, and geology) and also in non-earth science domains, for example, cadastral registration (van Oosterom and Lemmen 2001; van Oosterom et al. 2002). An information model must consider some specific concepts related to observation temporal locations and the used time reference system. Generally, both date and time and ordinal systems must be supported (Nativi and Ross 2007). It must also trade off between the unpractical requirement of precisely referring data to time and the unreasonable assumption that time is of little importance for spatial-information management and discovery.

Some position papers on this topic (Nativi et al. 2004; Nativi and Ross 2007; Tandy et al. 2006; INTERO 2006) list a number of properties of temporal information to be considered:

- Temporal knowledge may be relative, and either the relative times are more precisely known than absolute times or they cannot be described by a date (e.g. geological eras).
- Some temporal reference systems can have a moving reference time (e.g. weather forecasts are relative to the issue time, and current weather is "now").
- Temporal knowledge may be uncertain, that is, the exact relationship between two times may not be known precisely.
- Temporal information may be ordinal, where the periods are in sequence and are named (e.g. spring, summer, autumn, winter) and the transitions may not have specific dates.
- The granularity of temporal information may vary significantly (e.g. geological eras vs. the duration of a solar eclipse).

The earth sciences pose this challenge to spatial infrastructure initiatives. A valuable case in point is the INSPIRE initiative. Although the INSPIRE Directive (INSPIRE) has no mandatory requirements for temporal metadata in discovery, the INSPIRE working group on metadata has recognized requirements for temporal information that could be a factor in searching for data. Data for some themes (e.g. atmosphere, meteorology and oceanography) are fundamentally temporal and are commonly organized by date and time. A pilot study on discovery of data through temporal elements has been proposed (INSPIRE DT MD).

5.5   THE "COVERAGES" VIEW IN EARTH SCIENCE

Most earth science data are based on a composite approach, whereas most land management information is organized according to a georelational (or boundary) approach (Nativi et al. 2005). The composite approach is in essence a "bottom-up" means of organizing data, proceeding from individual measurement values to aggregated entities made up of those measurement values. On the other hand, the georelational approach to data organization is "top down," proceeding from meaningful aggregation entities to their actual measurements content (Molenaar 1991).

Present spatial information systems work with two fundamental spatial data elements: vectors and rasters. These data types have led to the more general concepts of features and coverages. Features generally represent geometric entities on the surface of the Earth (e.g., rivers, streets, lakes, and tracts of land). The characteristics of those entities are usually stored in a Data Base Management System (DBMS). Coverages, however, can be used to map composite data of the sort found in satellite images, radar observations or the output of forecast models.

In fact, in spatial information modelling, a coverage is a special case of feature (i.e. a subtype). It is defined as a feature that acts as a function to return values from its range for any direct position within its spatiotemporal domain (ISO 19123).

It is important to note that the information modelling approach of ISO TC211 enables the hard distinction between traditional vector and raster data types to be loosened. Indeed a coverage may be defined over any type of geometric domain (points, curves, Triangular Irregular Networks [TINs], solids, etc.)—not only grids.

An important coverage subtype is imagery. In terms of volume, imagery is the dominant form of spatial information. There will be required advanced semantic processing within the SII to automatically manage and process such huge amounts of growing data in order to extract valuable and useful information for decision support (e.g. automatic detection of features; data mining based on geographic concepts, knowledge extraction, etc.).

5.6   AN "OBSERVATIONS" MODEL FOR EARTH SCIENCE

Earth science data sets are generated primarily by observations of phenomena: data sets are used to capture and represent information related to complex earth science phenomena. In this context, a phenomenon may be considered as the property of one or more feature types of the earth system, the value of which is estimated by application of some procedure in an observation (Cox 2007a).

The schema depicted in figure 5.6 provides a schema of the described earth science observation model. The model is in line with the Observations and Measurements model proposed by OGC (Cox 2007a).

The feature of interest of an observation may be any feature having properties whose values are discovered by observation. In general, this will be an instance of a type from a catalogue representing the application domain for an investigation (Cox 2007a).

If the type of the *feature of interest* allows for a property (phenomenon) that varies temporally or spatially, then the value of the property is a coverage whose domain is the spatiotemporal extent of the *feature of interest*. Thus, the value of a corresponding observation result must also be a coverage. Generally, the *observation* domain may be different from the *feature of interest* domain (i.e. the observation does not cover the entire phenomenon domain or cover more than the actual phenomenon domain). Therefore, the feature of interest of this type of observation may be considered the spatial and temporal domain of a Sampling Feature (Cox 2007b), or proximate feature of interest.

The feature types defined in the Climate Science Modelling Language have been designed also to conform to the Observations and Measurements model (CSML UM); see figure 5.7 for the abstract CSML model—concrete CSML feature types specialize this model. CSML core feature types reflect various significant observational sampling regimes and geometries, for example, RaggedSection, ProfileSeries, Trajectory, GridSeries, etc. Each of these feature types may be regarded as an Observation sampling *feature of interest*. Furthermore, each CSML feature is associated with a CSML coverage class having a suitable spatiotemporal domain. These coverage classes may be regarded as the Observation result.

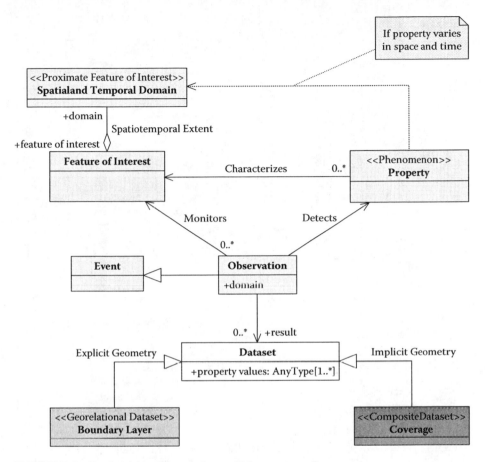

FIGURE 5.6 Earth science Observation and Measurement framework.

The geometry and topology of observation sets are a fundamental determinant of the scientific uses to which they may be put. It is no surprise that various information models in the earth sciences reflect these sampling properties (Tandy 2007), for example, the netCDF Common Data Model, the CSML feature types and the OGC Sampling Feature classes (Cox 2007b).

The sampling strategies in some cases may be regarded as "containers" for higher-level scientific information. Application users would like to search, discover and browse information by content rather than by container description. Therefore, advanced SII should adopt a harmonization framework that encompasses observation and measurements, feature and coverage information models. In effect, they are three different facets of the same real-world phenomenon. These three views are used to capture and encode different levels of explicitness of observed information content.

5.7　A SEMANTIC SII: BENEFITS TO EARTH SCIENCE

The potential benefits of a rich semantic SII to earth science are considerable. We consider two use cases that illustrate these potential benefits.

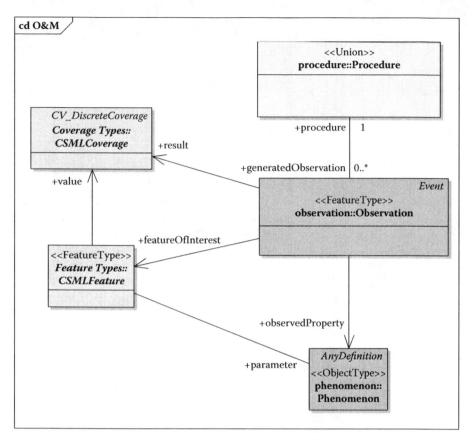

FIGURE 5.7   CSML feature types and the Observations and Measurements pattern.

First, a user may wish to undertake an on-demand flood-risk assessment (figure 5.8). The required components could all be available independently through the SII—daily analyzed rainfall data, high-resolution digital elevation models for the region of interest, and hydrological models available as geoprocessing services. These resources would need first to be discovered by searching against data set and service catalogues. The rainfall measurements might need to be analyzed into a gridded field for input to the hydrological model. A geoprocessing chain would have to be constructed and invoked to undertake these operations in the correct order, perhaps supplemented with additional information elements available locally (e.g. soil moisture data, etc.). The Business Process Execution Language (BPEL) (OASIS-BPEL) has been tested as a means to undertake such composition of data and service resources within SII applications (for instance, the OGC Web Services testbed exercises).

Another earth science-based example (figure 5.9) is an oceanographic researcher wishing to compare measurements of a mesoscale eddy from a marine field campaign with simulation data in order to study the detailed dynamics. This is a particular example of a very common problem in earth science, the need to validate simulation data (e.g. as output by the GMES Marine Core Service for operational ocean forecasting) against real-world observations. The SII can play a major role in

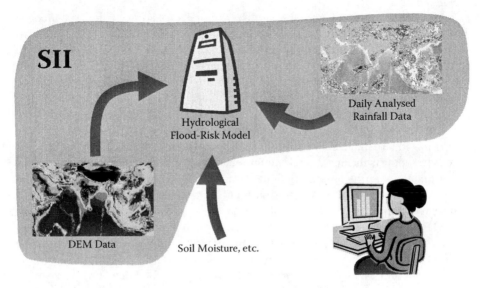

FIGURE 5.8   (See color insert.) SII use case: on-demand flood-risk modelling.

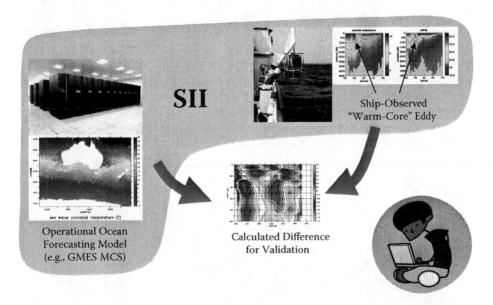

FIGURE 5.9   (See color insert.) SII use case: model validation against field campaign.

discovering suitable real-world validation data, extracting simulation "observation-equivalents" and portraying the difference fields.

For the above scenarios to be realizable, the SII needs to incorporate rich information models, and to make them available in formal catalogues. The feature catalogue provides such a "semantics repository" within the SII. Advanced conformance levels to the standard (ISO 19110) enable both inheritance and operations to be modelled.

Modelling inheritance enables reuse and specialization of domain models. For instance, a generic "tidal time-series" feature class could be declared (e.g. as in the IHO S-100 specification; Alexander et al. 2007). This could be extended in application schemas to new feature types appropriate for more specialized applications (figure 5.10), for instance by adding attributes for instrument or station details.

Similarly, modelling the behavior of feature types in a feature catalogue through their operations enables service chains to be constructed and advanced scientific workflows to be established. For instance, declaring that a "rainfall time-series" feature type supports the operation "calcSummerMean()" to calculate the average summer rainfall opens the possibility to discover services within the SII that implement the operation, and thereby to match a rainfall data source to an appropriate processing service (figure 5.11).

These two properties can be combined to obtain powerful semantic functionality. By declaring operations on base feature types, an implementing service is able to implement such operations for any derived feature type. For instance, a generic spatiotemporal coverage class could declare operations for calculating averages in space and/or time, and also subsetting in various ways (e.g., extracting a vertical "profile" of measurements through a 3D domain volume representing the atmosphere or ocean). A service could implement these operations and could be invoked on any subclass of the base coverage feature type.

Semantic service annotation (e.g. though OWL-S) will provide even more sophisticated functionality. Recent work (the Ontology Definition Metamodel developed by OMG; see http://www.omg.org/ontology/) has attempted to define mappings at a meta-level between knowledge representation systems (such as OWL) and the object modelling approaches (UML) used in SII. It may be that conceptual modelling (e.g. through UML) and formal ontologies are usefully regarded as dual approaches; however, the relationship between these two approaches to encapsulating semantics needs further exploration.

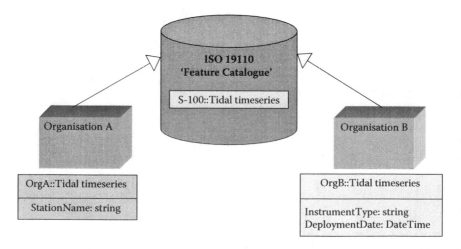

FIGURE 5.10   Feature catalogue: feature-type inheritance for reuse.

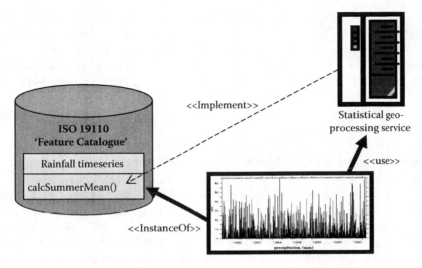

FIGURE 5.11 Feature catalogue: feature operations to support service chaining.

5.8   EARTH SCIENCE SII EXAMPLES: BIODIVERSITY AND CLIMATE CHANGE

Earth and space science informatics (ESSI) are not limited to data systems. Today, the focus is on infrastructures based on service and information technologies; they are called e-infrastructures or cyber-infrastructures; they are model driven. Interoperability enabled by spatial information and services standardization is the key factor of these systems.

The earth science community is already dealing with topics such as geoscience services, earth science-grid platforms, earth science information and service interoperability, etc. In fact, the community has been developing geosciences information infrastructures. Many earth science disciplines have been investigating specific information infrastructures, such as climate science, for example, LAS (Live Access Server) (LAS site), OPeNDAP/THREDDS (Open-source Project for a Network Data Access Protocol/Thematic Realtime Environmental Distributed Data Services) (OPeNDAP site) (THREDDS site); biodiversity, for example, GBIF (Global Biodiversity Information Facility) (GBIF portal), OBIS (Ocean Biogeographic Information System) (OBIS portal); weather and forecast, for example, WIS (WMO Information System) (WMO WIS page); hydrology, for example, HIS (Hydrological Information System) (HIS site); etc.

There is a clear need to start from these information community infrastructures to make them fully and effectively interoperable. In effect, decision makers, researchers and scientists of many societal benefit areas (SBA) urge the earth science community to conceive these infrastructures in a coordinated and multidisciplinary context. There exists the necessity to lead systematic investigations, considering requirements coming from important SBA and thoroughly discussing solutions with the entire earth science community. Certainly, all earth science discipline infrastructures will benefit from a reengineering process around spatial information standards. In fact,

SII could play a decisive role in this process, providing the reference infrastructure and, thus, enabling the integration of the various earth science domain solutions.

Internationally, some important initiatives and programs are already active in the investigation of such a multidisciplinary framework, among them the European GMES initiative with the ESA GMES Service Elements (GSE) program, the international GEOSS along with the IEOS (Integrated Earth Observing System) and GOOS (Global Ocean Observing System), the U.S. NASA Geo-sciences Interoperability Office (GIO) program, the ESA Service Support Environment (SSE), the U.S. NSF GEON (Geosciences Network) project and the international Geo-Grid project. Soon, other valuable initiatives will be launched to address other SBA interested in the location-based services offered by Galileo.

A valuable investigation to address important SBA is represented by the experiments that interest the climatology and biodiversity communities and their related infrastructures' interoperability.

Biodiversity is a handy, one-word name for all the species on the Earth, the genetic variety they possess and the ecological systems in which they participate. Another way of thinking about biodiversity is as the "living resources" portion of "natural resources." A large part of the primary data on biodiversity is the 1.5 to 2.0 billion specimens held in natural history collections, as well as many geographical and ecological observations recorded by various means and stored in various media (GBIF portal).

Clearly, it is of paramount importance to openly share and put to use vast quantities of global biodiversity data advancing scientific research in many disciplines, promoting technological and sustainable development, facilitating the conservation of biodiversity and the equitable sharing of its benefits, and enhancing the quality of life of members of society. In making living resource policy and management choices, decision makers should be able to discover, access and use biodiversity primary information, in connection with other important spatial information layers (MOU 2006). To facilitate this process the GBIF was established in 2001. It supports policy- and decision-makers, research scientists and the general public all around the world to electronically access the world's supply of primary scientific data on biodiversity. GBIF is an international organization established by the Memorandum of Understanding for the Global Biodiversity Information Facility. Today, 40 countries and 33 international organizations are participants in GBIF. It now makes 120 million individual records available from 1,000 databases on 200 servers in 30 countries. This data pool can be used for large-scale scientific questions that no single research group could hope to answer using its own data sets (Hannu 2007).

According to the Secretariat of the Convention on Biological Diversity, one of the biggest threats to biodiversity is climate change; the links between biodiversity and climate change run both ways: biodiversity is threatened by human-induced climate change, but biodiversity resources can also reduce the impacts of climate change on people and production. Hence, there exists a growing demand for accurate forecasting of the effects of global warming on biodiversity. Climate change threatens to commit 15% to 37% of species to extinction by 2050, accelerating a mass extinction precipitated by widespread land use changes (Hannu 2007). The need to assess these impacts and recommend solutions to policymakers is correspondingly

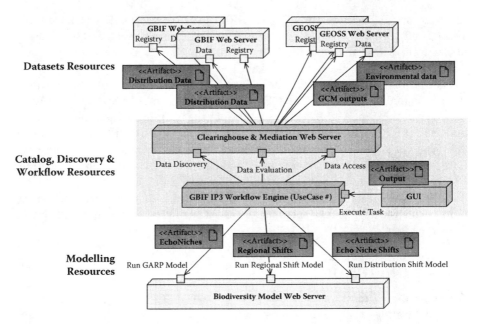

FIGURE 5.12 Architectural schema for the Biodiversity-Climate Change Interoperability Pilot.

acute. Such analyses must integrate enormous volumes of data from biodiversity archives, satellite remote sensing and climatic data. This is a task that requires a multidisciplinary science infrastructure. Therefore, there is the need for an independent and permanent information infrastructure to experiment and compare these multidisciplinary models.

In the framework of the GEOSS initiative, the GBIF Secretariat working with IEEE, WMO and other partners (i.e. the Italian National Research Council, University of Helsinki and University of Ottawa) will participate in a couple of interoperability experiments: the GEOSS Interoperability Process Pilot Projects (IP3) (Khalsa et al. 2007) and the GEOSS Architecture Implementation Pilot (GEO 2007). This investigation includes the development of formal scenarios and use cases that employ GBIF data to address climate change issues. These interoperability pilots will make use of the present GBIF global information infrastructure for sharing, accessing and using biodiversity data. This infrastructure builds on contributions of a large number of member countries and organizations, especially the work of the open Biodiversity Information Standards body, TDWG (TDWG Web site).

The high-level component architecture of the GBIF interoperability pilot is depicted in figure 5.12.

The pilot architecture makes use of a biodiversity model component: an Open-Modeller server implementation. OpenModeller is an innovative framework for carrying out various computer modelling tasks using different algorithms in an integrated but open manner (Santana et al. 2006).

The system aims to enhance the present OpenModeller workbench with GEOSS metadata and online access to new GBIF data access services based on a resource-oriented architecture (i.e. the GBIF ReST services) (GBIF ReST tutorial Web site). In order to discover and access climate, environmental and biodiversity layers from within the OpenModeller installation, a new computing server was developed; it makes use of a new AJAX-based interface and an ISO 19115 metadata crosswalk to discover and access the desired climate change data layers from remote sources registered at the GEOSS Clearinghouse. AJAX stands for Asynchronous JavaScript And XML; it is a type of programming—based on JavaScript and HTTP requests—made popular in 2005 by Google (with Google Suggest); AJAX is not a new programming language, but a new way to use existing standards; it is used to create better, faster and more user-friendly Web applications (W3C AJAX tutorial).

These investigations will provide system requirements for the GEOSS Clearinghouse, which will be an interoperable register of registries, including the GBIF data registry system.

5.9   CONCLUSIONS

Although the earth sciences may be considered an application domain for SII, there is nevertheless a significant synergy between the two—both are focused on resource integration, have a fully global scope and are multidisciplinary by nature. This is reflected by the parallel developments of informatics and grid infrastructures in the earth sciences, and in the key SII driver of environmental policymaking.

There are a number of particular benefits that will flow to SII through application in the earth sciences. The temporal nature of earth-related phenomena requires a much more sophisticated approach to modelling time (for both data and metadata) than is required in other fields. The "coverage" view of data as a field over some spatiotemporal domain is ubiquitous in the earth sciences. Finally, the process of observation is the primary means of obtaining earth science data, and has useful characteristics that can be captured through a generic "Observations model," incorporating aspects of both the observation process and the sampling regime.

REFERENCES

Alexander, L., M. Brown, B. Greenslade, and A. Pharaoh. 2007. Development of S-100: The New IHO Geospatial Standard for Hydrographic Data. http://www.iho.shom.fr//COMMITTEES/CHRIS/TSMAD/S-100_Info_Paper.pdf.

Biosphere Web site. http://www2.sims.berkeley.edu/academics/final-projects/biosphere/eimore.html.

CSML UM, NERC *DataGrid, Climate Science Modelling Language version 2 Users Manual* (available at http://ndg.nerc.ac.uk/csml/).

Cox, S., ed. 2007a. Observations and Measurements. Part 1. Observation Schema. Open Geospatial Consortium document 07-022r1.

Cox, S., ed. 2007b. Observations and Measurements. Part 2. Sampling Features. Open Geospatial Consortium document 07-002r3.

CYCLOPS. 2006. Technical Annex, Cyber-Infrastructure for Civil Protection Operative Procedures (CYCLOPS) SSA project N. 031874, March 2006.

CYCLOPS Web site. http://www.cyclops-project.eu/.

DDC Web site. http://www.ipcc-data.org.

EGEE Web site. http://www.eu-egee.org/.

Foster, I., and C. Kesselman. 1999. *The Grid: Blueprint for a New Computing Infrastructure.* San Francisco, CA: Morgan Kaufmann.

Foster, I., and C. Kesselman. 2006. Scaling System-Level Science: Scientific Exploration and IT Implications. *IEEE Computer* 39 (11):31–39.

Foster, I., C. Kesselman, and S. Tuecke. 2001. The Anatomy of the Grid: Enabling Scalable Virtual Organizations. *Int. J. High Performance Computing Applications* 15 (3):200–222.

Foster, I., C. Kesselman, J. M. Nick, and S. Tuecke. 2007. The Physiology of the Grid: An Open Grid Services Architecture for Distributed Systems Integration. http://www.globus.org/research/papers/ogsa.pdf, 24th September 2007.

GBIF portal http://www.gbif.org.

GBIF ReST tutorial site: http://data.gbif.org/tutorial/services.

GEO. 2007. Architecture Implementation Pilot Call for Participation (CFP). Group on Earth Observations. http://www.earthobservations.org/docs/CFP_GEOSS_AR-07-02-114.2007.pdf.

GeoSciML Web site. https://www.seegrid.csiro.au/twiki/bin/view/CGIModel/GeoSciML.

Hannu, S. 2007. GBIF Pilot, Response to the GEOSS Architecture Implementation Pilot Call for Participation.

HIS web site. http://www.cuahsi.org/his/.

INSPIRE Directive 2007/2/EC of the Parliament and of the Council, http://eur-lex.europa. Eu/JOHtml.do?uri-OS:L:2007:108:SOM:EN:HTML.

INSPIRE. 2007. Metadata DT, Temporal Reference Metadata for Discovery: Pilot Study Proposal. INSPIRE document. EC Joint Research Center, Ispra, Italy.

INTERO. 2006. Temporal Extent Metadata Implementation Model. ed. S. Nativi. Italian National Earth & Environment Research Community (INTERO) document for INSPIRE.

ISO. 2000. ISO/TR 19121:2000, Geographic Information—Imagery and Gridded Data. Geneva: International Organization for Standardization.

ISO. 2002a. ISO 19101:2002, Geographic Information—Reference Model. Geneva: International Organization for Standardization.

ISO. 2002b. ISO 19108:2002, Geographic Information—Temporal Schema. Geneva: International Organization for Standardization.

ISO. 2003a. ISO 19111:2003, Geographic Information—Spatial Referencing by Coordinates. Geneva: International Organization for Standardization.

ISO. 2003b. ISO 19115:2003, Geographic Information—Metadata. Geneva: International Organization for Standardization.

ISO. 2005a. ISO 19103:2005, Geographic Information—Conceptual Schema Language. Geneva: International Organization for Standardization.

ISO. 2005b. ISO 19109:2005, Geographic Information—Rules for Application Schema. Geneva: International Organization for Standardization.

ISO. 2005c. ISO 19110:2005, Geographic Information—Methodology for Feature Cataloguing. Geneva: International Organization for Standardization.

ISO. 2005d. ISO 19118:2005, Geographic Information—Encoding. Geneva: International Organization for Standardization.

ISO. 2005e. ISO 19123:2005, Geographic Information—Schema for Coverage Geometry and Functions. Geneva: International Organization for Standardization.

ISO. in press. ISO 19101-2, ISO/PDTS 19101-2, Geographic Information—Reference Model—Part 2: Imagery. Geneva: International Organization for Standardization.

ISO. in press. ISO/DIS 19136, Geographic Information—Geography Markup Language (GML). Geneva: International Organization for Standardization.

ISO. n.d. ISO/PDTS 19129, Geographic Information—Imagery, Gridded and Coverage Data Framework. Unpublished document. Geneva: International Organization for Standardization.

Khalsa, S. Nativi, T. Ahern, R. Shibasaki, and D. Thomas. 2007. The GEOSS Interoperability Process Pilot Project. Paper presented at IGARSS07, Barcelona, July 2007.

LAS Web site. http://ferret.wrc.noaa.gov/ferret/LAS/.

Lawrence, B., R. Cramer, M. Gutierrez, K. Kleese van Dam, S. Kondapalli, S. Latham, R. Lowry, K. ONeill, and A. Woolf. 2004. The NERC DataGrid: Googling Secure Data. UK e-Science Programme All Hands Meeting (AHM2004), Nottingham, U.K., August 31–September 3.

Lowe, D., A. Woolf, B. Lawrence, P. Cooper, and R. Lott. 2006. Standards Challenges in the Climate Sciences. AGU Fall Meeting, San Diego, CA, December 2006. *Eos Trans. AGU* 87 (52), Fall Meet. Suppl., Abstract IN43C-0916.

Molenaar, M. 1991. Status and Problems of Geographical Information Systems. The Necessity of a Geo-Information Theory. *ISPRS Journal Remote Sensing* 46:85–103.

MOU. 2006. Memorandum of Understanding for the Global Biodiversity Information Facility, approved at GB12 in Cape Town, South Africa, April 2006.

Nativi, S., B. Blumenthal, T. Habermann, D. Hertzmann, R. Raskin, J. Caron, B. Domenico, Y. Ho, and J. Weber. 2004. Differences Among the Data Models Used by the Geographic Information Systems and Atmospheric Science Communities. Proceedings of AMS—20th IIPS Conference, Seattle, WA, January 2004.

Nativi, S., J. Caron, E. Davis, And B. Domenico. 2005. Design and Implementation of netCDF Markup Language (NcML) and its GML-Based Extension (NcML-GML), *Computers & Geosciences Journal* 31 (9): 1104–1118.

Nativi, S., and G. Ross. 2007. Temporal Extension Metadata: Position Paper. INSPIRE DT Metadata document. EC Joint Research Centre, Ispra, Italy.

NDG Web site. http://ndg.nerc.ac.uk.

NERC science topics. http://www.nerc.ac.uk/funding/application/topics.asp.

OBIS portal. http://www.iobis.org/.

OPeNDAP Web site. http://www.opendap.org/.

OASIS-BPEL: Web Services Business Process Execution Language Version 2.0. http://docs.oasis-open.org/wsbpel/2.0/OS/wsbpel-v2.0-OS.html.

Santana, F. S., R. R. Fonseca, A. M. Saraiva, P. L. P. Corrêa, C. Bravo, and R. Giovanni. 2006. OpenModeller—An Open Framework for Ecological Niche Modeling: Analysis and Future Improvements. Paper presented at the 2006 World Conference on Computers in Agriculture. Orlando, FL. http://www.WCCA.2006.org.

Tandy, J. 2007. Moving Forward from the RAL Feature Workshop. Paper presented at the GO-ESSP Community Workshop, Université Pierre et Marie Curie (Jussieu), Paris, France, June 11–13, 2007.

Tandy J., B. Lawrence, and G. Ross. 2006. Explanation of Times for Observation and Simulation. MetOffice document for INSPIRE.

TDWG Web site. http://www.tdwg.org.

THREDDS Web site. http://www.unidata.ucar.edu/projects/THREDDS/.

TeraGrid Web site. http://www.teragrid.org/.

van Oosterom, P. J. M. and C. H. J. Lemmen. 2001. Spatial Data Management on a Very Large Cadastral Database. *Computers, Environment and Urban Systems* 25 (4/5):509–528.

van Oosterom, P. J. M., B. Maessen, and C. W. Quak. 2002. Generic Query Tool for Spatio-Temporal Data. *International Journal of Geographical Information Science*. 16 (8):713–748.

W3C AJAX tutorial. http://www.w3schools.com/ajax/default.asp.

WMO WIS Web page. http://www.wmo.ch/pages/themes/wis/index_en.html.

Woolf, A., R. Cramer, M. Gutierrez, K. Kleese van Dam, S. Kondapalli, S. Latham, B. Lawrence, R. Lowry, and K. O'Neill. 2005. Standards-Based Data Interoperability for the Climate Sciences. *Meterological Applications* 12 (1):9–22.

Woolf, A., B. Lawrence, R. Lowry, K. Kleese van Dam, R. Cramer, M. Gutierrez, S. Kondapalli, S. Latham, K. O'Neill, and A. Stephens. 2004. Integrating Distributed Climate Data Resources: The NERC DataGrid. In *Use of High Performance Computing in Meteorology*, ed. Walter Zwiefelhofer and George Mozdzynski. Singapore: World Scientific.

Woolf, A., B. Lawrence, R. Lowry, K. Kleese van Dam, R. Cramer, M. Gutierrez, S. Kondapalli, S. Latham, D. Lowe, K. O'Neill, and A. Stephens. 2006. Data Integration with the Climate Science Modelling Language. *Advances in Geosciences* 8 (1):83–90.

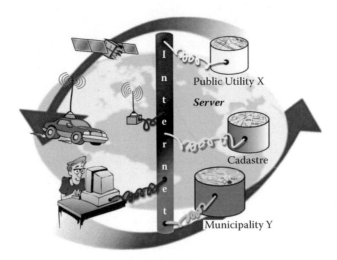

FIGURE 0.1

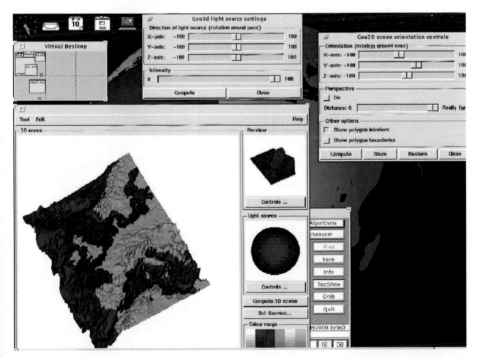

FIGURE 0.2

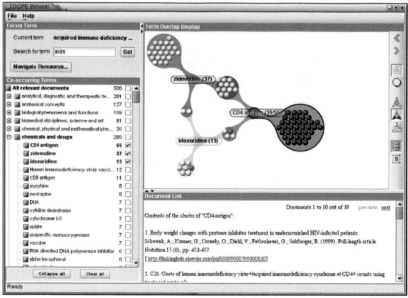

FIGURE 3.1

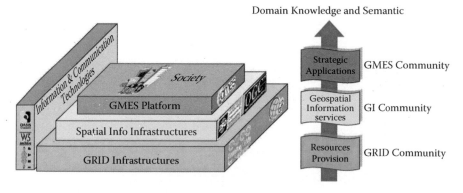

FIGURE 5.2

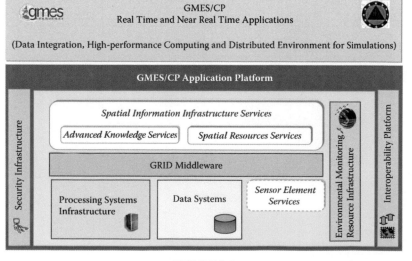

FIGURE 5.3

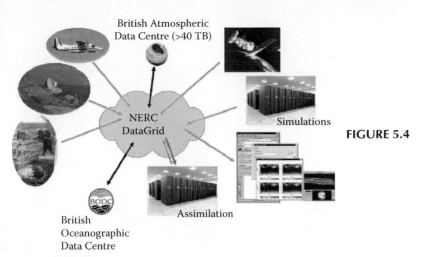

British Atmospheric
Data Centre (>40 TB)

NERC
DataGrid

Simulations

British
Oceanographic
Data Centre

Assimilation

FIGURE 5.4

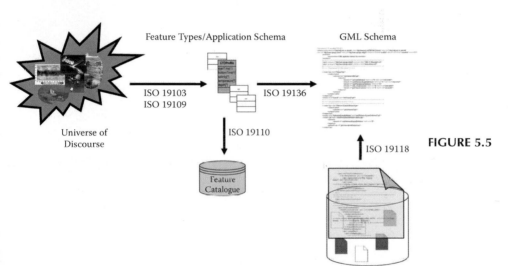

Feature Types/Application Schema

GML Schema

ISO 19103
ISO 19109

ISO 19136

Universe of
Discourse

ISO 19110

ISO 19118

FIGURE 5.5

Feature
Catalogue

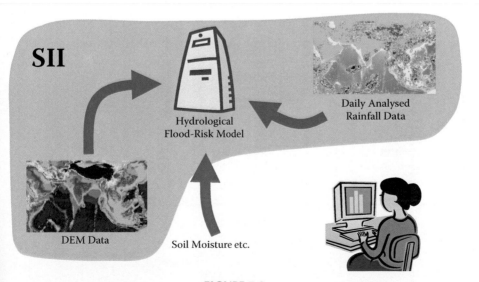

SII

Hydrological
Flood-Risk Model

Daily Analysed
Rainfall Data

DEM Data

Soil Moisture etc.

FIGURE 5.8

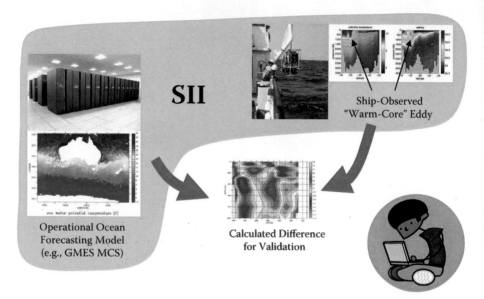

FIGURE 5.9

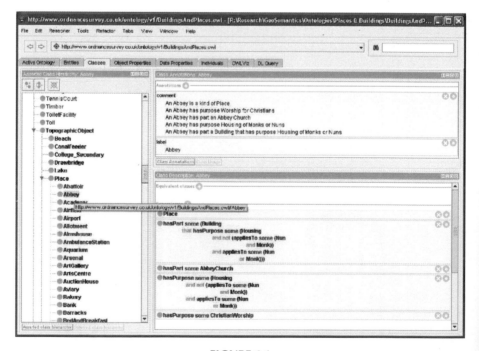

FIGURE 6.1

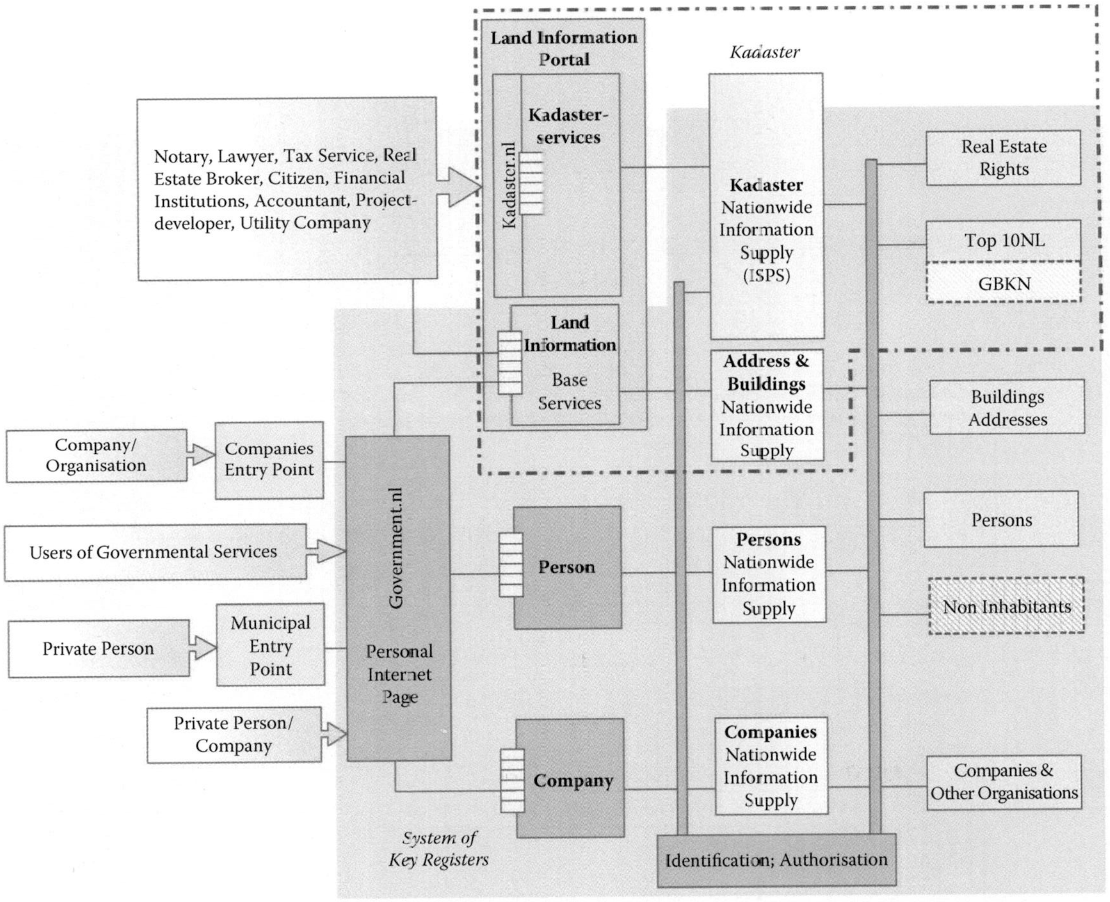
Land Information Portal
Kadaster-services
Kadaster.nl
Land Information
Base Services
Kadaster
Kadaster Nationwide Information Supply (ISPS)
Address & Buildings Nationwide Information Supply
Persons Nationwide Information Supply
Companies Nationwide Information Supply
Real Estate Rights
Top 10NL
GBKN
Buildings Addresses
Persons
Non Inhabitants
Companies & Other Organisations
Notary, Lawyer, Tax Service, Real Estate Broker, Citizen, Financial Institutions, Accountant, Project-developer, Utility Company
Company/ Organisation
Companies Entry Point
Users of Governmental Services
Private Person
Municipal Entry Point
Private Person/ Company
Government.nl
Personal Internet Page
Person
Company
System of Key Registers
Identification; Authorisation

FIGURE 9.1

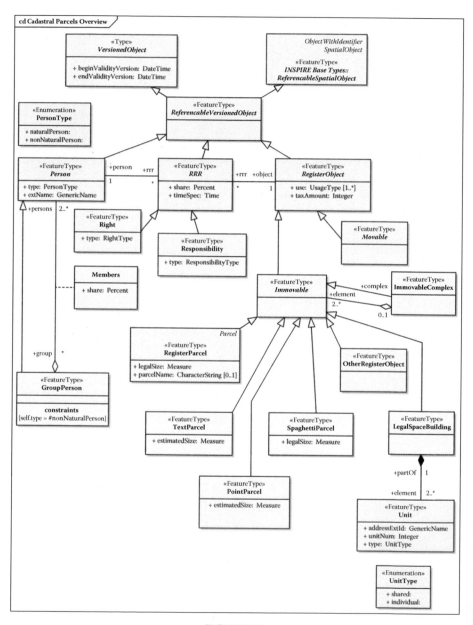

FIGURE 9.3

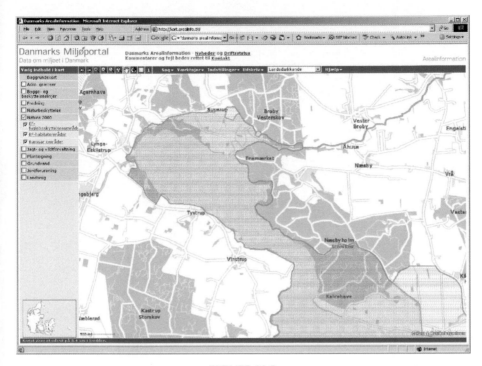

FIGURE 11.2

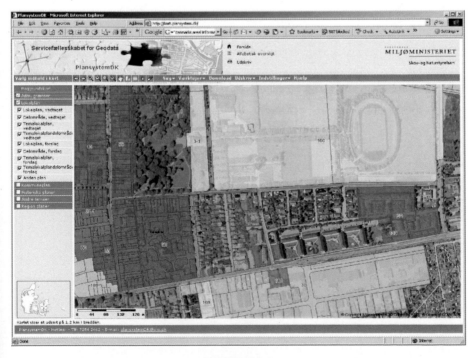

FIGURE 11.4

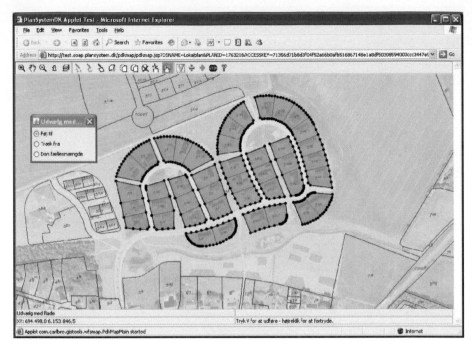

FIGURE 11.6

6 Opportunities and Challenges in Exploiting Semantics as an Aid to Information Integration

*A National Mapping Agency Perspective**

Catherine Dolbear and Glen Hart

CONTENTS

This chapter describes how Ordnance Survey is promoting the use of geographic information (GI) within the information economy through the use of Semantic Web technologies. It details the importance of GI to the general task of information integration and how the development of spatial information infrastructures that are reliant on standards alone may limit the adoption of GI by the wider information industry. We believe that Semantic Web technologies can help to overcome the limitations of a standards-only approach, though we also highlight the challenges that are faced in the development of this immature technology. Lastly, our chapter describes the research currently being conducted to overcome some of these limitations.

---

* © Crown Copyright 2007 Reproduced by permission of Ordnance Survey. This chapter has been prepared for information purposes only. It is not designed to constitute definitive advice on the topics covered and any reliance placed on the contents of this chapter is at the sole risk of the reader.

This chapter presents our research into semantic data integration, motivated by the desire to increase the use of geographic information within Britain. We discuss our work on authoring topographic ontologies, and how they are being used for semantic data integration. This chapter also outlines the options for extending semantic technologies to allow for spatial reasoning, and describes how we are using data ontologies to bridge between our spatial database and domain-level ontologies. Finally, with our experimental Resource Description Framework (RDF) product, we provide one future direction for semantic data products.

6.1   GROWING THE GEOGRAPHIC INFORMATION ECONOMY

As the national mapping agency for Great Britain, Ordnance Survey has a responsibility not only to provide a topographic referencing framework but also to increase the use of GI within the country. Given that the purpose of a spatial information infrastructure is to aid the exchange and integration of GI, it follows that the creation of such an infrastructure should aid Ordnance Survey's objective of spreading the use of GI. However, this can only be achieved if the manner in which the infrastructure is constructed does not inhibit its adoption by the non-GI specialist.

Using geographic information raises many of the same issues as using other types of information. However, there are some problems specific to GI that also need to be considered (Hart and Dolbear 2007), the most important of which are the bridging role played by GI across multiple domains and the complexity of spatial data. These differences need to be understood while not being overplayed—the growth of GI is closely linked to the growth of the wider information industry: grow the overall information market and the GI market will also grow. So, one might also ask, how can GI help to grow this larger market?

One of the most important technical challenges faced by the information industry is how to integrate disparate and varied information sources together as part of an overall business process. The challenge this presents is enormous and is consistently underestimated, because the focus is usually on software and hardware integration, rather than a deep understanding of how information impacts business operations. Information systems that contain GI are often able to exploit the GI component to aid the integration process. However, an information system rarely exists to manipulate geographic information alone. For example, an environmental agency may be interested in where animal habitats are located, but the "where" is rarely the most important factor. Rather, the agency's primary concern will be the nature of the habitats, how threatened they are and so on. Similarly, an insurance company may be interested in location to evaluate insurance risk and for customer marketing purposes, but its main focus will be the items insured, the overall risk and the values involved. The role of GI is that of the bridesmaid and not the bride: always supporting, never the real focus. However, location is commonly a theme that is shared by disparate information sources, although its form may vary: location may be represented in the form of explicit coordinates, postal addresses, topographic identifiers and so on. As a result, geographic information can perform an essential role—acting as a common factor linking these information sources and assisting information integration. This role is generally recognized and indeed encouraged by the "GI industry," as demonstrated by European initiatives such as INSPIRE (see chapter 1) and national initiatives such as the Digital National Framework (see http://www.dnf.org/Pages/home/default.asp).

One way in which geographic information does differ from other digital information sources, however, is its complexity, which has limited its adoption outside of the GI community. In recent years, however, the take-up of GI by the wider information community has been growing, as mainstream information technology players such as Oracle have included spatial capabilities in their basic product offerings. Despite these positive developments, a major impediment to the full exploitation of GI has been the difficulty of integrating geographic information with other sources and embedding it into business processes.

The complexity of GI is related to the structural nature of the data itself (the need for spatial indexing, the representation of geometry and topology, etc.), the great variety of forms that it may take and the variety of sources that generate it. To a degree, standards can help, but they work best within single communities, and indeed standards may themselves impede the use of data across communities where different standards apply. For pragmatic reasons, standards are often limited to specification of structure (the information syntax) rather than the semantics. GML (GML 2004) is very effective within the GI world, but essentially it standardizes geometric and topologic syntax. It does not concern itself with the semantic content that it represents, nor is it well known outside the GI community. Similarly, infrastructures aimed at meeting the needs of individual communities, while promoting the exchange of information within that one community, can impede exchange across disparate communities. Given that GI often acts as the fabric upon which other information can be integrated, the development of a spatial information infrastructure must be carefully crafted to ensure that it does not impede the use of GI by the much larger information industry. We are not making the argument that the use of standards should be avoided within a spatial information infrastructure; rather, we are warning that we cannot rely on standards alone to solve the wider issue of interoperability across different communities.

Therefore, an important element of the information integration solution lies beyond standards and syntax. Semantic diversity is not something to be engineered away, but is an important aspect of any solution. The trick is not to create a single worldview to which everyone must conform, but to create the means by which to navigate though multiple worldviews. This places the semantics of these worldviews at the heart of any solution, and indeed this paradigm of decentralization is a cornerstone of the Semantic Web.

The Semantic Web (Berners-Lee, Hendler, and Lassila 2001) encodes such semantics through the explicit description of information in *ontologies* and Semantic Web technology being developed today will offer the means to move between these different descriptions. An ontology encodes the meaning of the concepts in a particular domain by detailing the relationships between the concepts. Data in a database can be seen as instances of these ontology concepts. Furthermore, because the ontology can be machine-processed, reasoning can be performed on the concepts and, by extension, on the data that they describe.

Commenting on the future of the Semantic Web, Tim Berners-Lee asserted that, "We need to look at existing databases and the data in them" (Runciman 2006) and harness the Semantic Web for data integration purposes. In fact semantic technology can also address the problem of the "semantic gap": the difference between how a domain is viewed in its full richness at a business level, and how the objects within the domain are represented in a much cruder form, with fewer categories or descriptions, within a database. For example, Ordnance Survey requires surveyors to capture "real-world objects" such as houses, warehouses, factories and so on. However, within the data, these different objects are simply identified as "buildings" (although a textual description may be also associated with the object). Using an ontological description of the real-world objects and by reasoning over the data, this information loss can be partially recovered. Although it may be possible to achieve the same result using a GIS or SQL query, the ontology enables the definitions to be explicit and visible rather than buried as code within an application. However, the Semantic Web is still in its infancy—there are many challenges ahead before it can become effective in this new role.

6.2   ORDNANCE SURVEY AND SEMANTIC WEB TECHNOLOGIES

The generalist nature of Ordnance Survey information and its inherent purpose as a referencing framework mean that it is a useful component in many applications that require information integration. The Geosemantics research group at Ordnance Survey is therefore experimenting with the use of semantic technologies to close the semantic gap between our worldview and the data that have been actually captured, and also to aid information integration. We are therefore conducting research in a number of areas:

*   How to author ontologies (in order to build a topographic domain ontology)
*   How to use Semantic Web technologies to aid the task of information integration
*   How to represent and manipulate spatial data within the Resource Description Framework (RDF; Manola and Miller 2004) with a view to produce information products that are maximally suited for integration

At present we are not looking into areas such as semantic translation but have identified this as a necessary area for future work.

6.2.1   TOPOGRAPHIC ONTOLOGY AND AUTHORING ONTOLOGIES

We are developing a topographic domain ontology to help bridge the "semantic gap" between the knowledge we hold about the topographic domain and the data as they are currently collected today. This ontology will provide explicit semantic descriptions of the features within our databases and enable additional classifications to be inferred. And, as with the information itself, these semantic descriptions are also generalist in nature. This means the ontology can act as a semantic referencing framework, enabling others to share and specialize the semantic descriptions for their own use, in the same way as the geographic information we currently supply is used and specialized.

The topographic ontology that Ordnance Survey is building is modular in nature, including modules on transport, buildings and places, and land cover. Figure 6.1 shows a Protégé screenshot of part of the Buildings and Places ontology.

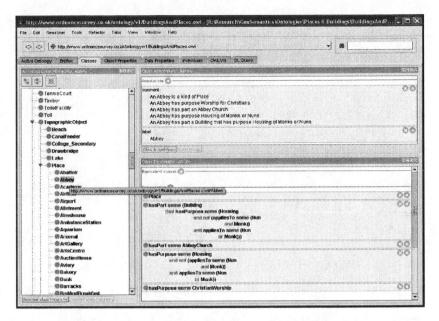

FIGURE 6.1 (See color insert.) Protégé screenshot of part of the Ordnance Survey Buildings and Places ontology.

Some of these reuse common concepts and relationships, such as mereology ("part-of" relationships), spatial relationships and language concepts like abbreviations or synonyms, so we have also started to identify these as separate modules in themselves, reused by our main domain ontologies. The point here is that the ontological approach avoids the traditional categorizing of geography, and indeed products, into rigid themes or layers. Although we at Ordnance Survey may find it convenient to divide the world in certain ways, to facilitate our own ontology building, our customers have their own worldviews and their own categories and concepts. Customers can pick and choose concepts from several different Ordnance Survey ontologies and add their own terminology or additional semantics in order to create their own ontologies and describe the combination of their information and ours. This vision goes beyond standards, allowing the free movement between different perspectives, rather than imposing a limited number of standardized ones.

To achieve this vision, we are tackling several technology problems, including how to author ontologies and how to join up semantic and spatial technologies, as well as issues surrounding the "semantic gap" and semantic reasoning with high volumes of information.

We have developed a methodology for authoring ontologies that combines two aspects. The first is the human-readable or "conceptual" aspect, which is written by one or more experts in the relevant domain. The second component is the computer-parsable or "computational" aspect, which is created by manual or automatic conversion of the conceptual ontology into a Semantic Web language such as the Web Ontology Language (OWL 2004). Our method is centered on the domain expert, because we believe it is important that control of the ontology and the authoring process remains with the human source of the knowledge: the domain expert. We have developed a structured natural language called Rabbit (Hart, Dolbear and Goodwin 2007) in which the domain expert can name concepts and relationships and specify axioms describing or defining the concepts in a way that makes sense to them. The Rabbit sentences then act as easy-to-understand documentation of the OWL, and it is at the Rabbit level of the ontology that domain experts determine whether to reuse or merge concepts from a third-party ontology with their own ontology's concepts. As shown in figure 6.1, the conceptual aspect of the ontology is written in Rabbit and includes sentences such as:

An Abbey is a kind of Church.

An Abbey has purpose Housing of Monks or Nuns.

These two sentences are represented in OWL as

Abbey ⊑ Church.

Abbey ⊑ hasPurpose ∃ Housing ⊔ appliesTo ∃ (Monk ⊓ Nun)

At present Rabbit sentences are "hand compiled" into OWL, there being well-defined mappings between the Rabbit structures and OWL. In conjunction with the University of Leeds we are developing Confluence, a plug-in to Protégé that will enable Rabbit to be used as the authoring language and then automatically converted to OWL. Once in OWL the ontology is available to any of the popular reasoners, such as FaCT++ and Pellet.

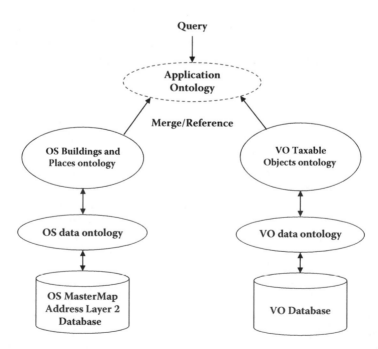

FIGURE 6.2   Prototype semantic data integration system.

At present we are merging ontologies either by simply importing the whole ontology or by the creation of an intermediate. The former case works well where the imported ontology is either entirely relevant to the recipient or where it is small in size. Where only a portion of a large ontology is required, then at present we are forced to either import the whole ontology or to copy a subset either directly into the recipient or as a separate module. Ideally we would like to be able to develop a means to import only those aspects of the ontology that are relevant to the recipient and we are looking at ways this may be achieved through either ignoring particular axioms or whole concepts/classes. Figure 6.2 shows a high-level view where two reference ontologies, the Ordnance Survey Buildings and Places and the Valuation Office's Taxable Object Ontology, are merged to form an application ontology (shown as a dotted ellipse) that enables cross-domain queries to be carried out. Ideally we would like the application ontology to contain those concepts and axioms specific to the application plus additional concepts referenced from the two reference ontologies. At present this can only be achieved by importing the whole ontologies. We are currently modifying Rabbit to enable concepts in a referenced ontology to be "switched off"—these concepts are visible but are not interpreted and Confluence would not pass any of their axioms on to OWL. By doing so, the view that the application ontology has of the reference ontology becomes restricted, enabling only the appropriate parts to be used.

6.2.2 SEMANTIC DATA INTEGRATION

To address the semantic interoperability problem, our intention is to develop a prototype demonstrator, shown in figure 6.2, that links two data sets: one from Ordnance Survey and one from a customer, for example, the Valuation Office (VO). Each of the data sets used within the prototype contains data collected according to some model of the world, using a set of terminology whose semantics are understood by that organization. We differentiate between *domain ontologies*, which are a formalization of the knowledge in a subject area (domain) such as topography, ecology, biology, etc., and *data ontologies*, which describe a data source and will include information about how it can be understood in terms of the domain terminology.

The two data sets will then be linked semantically through the merged ontology, and can then be queried as if they were one data source. For example, a query in the Ordnance Survey and Valuation Office data sets could be "Find me all addresses with a taxable value over £500,000 in Southampton." Although much has been said about the utility of semantics as an aid to data integration, we have not as yet come across any studies that quantify the advantages in practice. One aim of our prototype is therefore to understand how semantic technology adds value to the data integration process. Work in this area and with respect to topographic data integration has been conducted previously (Uitermark 2005) in a manner we believe is informative, and could form the basis for future research.

We are currently authoring a "Buildings and Places" domain ontology describing the knowledge held by Ordnance Survey about different Buildings (ranging from Animal House to Windmill, which are single structures) and Places (including Breweries and Vineyards, which are areas of land that may contain more than one structure). Similarly, a customer such as the Valuation Office has its own perspective on the world and its own terminology, and we intend to build a domain ontology to model this as well. This means that the data integration can take place between the two domain ontologies, using their semantic descriptions, rather than directly at the database level, where meaning may be hidden in the table structure. From this it is clear that semantic technology is no silver bullet to the data integration problem—considerable effort still needs to be invested in the building of domain ontologies, both by Ordnance Survey and its customers. However, this process of making organizational knowledge explicit only needs to be carried out once, as the domain ontology is reusable for any integration exercise. Furthermore, as we have found, it is incredibly useful in itself, as a way of identifying inconsistencies in granularity of the specification, and teasing out assumptions and organizational expectations that may not be apparent to the customer, or buried in hundreds of pages of text documentation.

6.2.3 EXTENDING SEMANTIC TECHNOLOGIES TO MEET THE NEEDS OF GI

Within the semantics community, there has been little attempt so far to address the problems of how to combine spatial and semantic reasoning over instance data. Significant work has been done to formalize spatial reasoning based on explicit topology such as the 9 Intersection Model (Sharif, Egenhofer, and Mark 1998) and RCC-8 (Randell, Cui, and Cohn 1992). However, a very significant proportion of all geospatial information does not contain explicit semantics describing topological information, as these are normally calculated on the fly using the geometry. Therefore, we are beginning to look into how semantic query languages like SPARQL (Prud'hommeaux and Seaborne 2006) can be extended to incorporate spatial elements. Our experiments so far have raised questions about how well the technologies scale and the inability of reasoners to act on implicit spatial relationships that need to be computed. Our efforts are currently concentrated on finding ways around the former issue.

As well as spatial and semantic queries, another significant stumbling block is the question of how to link an ontology to a database. Currently there are three main options. One is to convert semantic queries into SQL queries, wrapping the results as a virtual RDF graph. For example, the SQL query might be "SELECT FID FROM TopographicArea WHERE Theme = 'Building';" which will return all features in the database table "TopographicArea," where the Theme column is "Building." Each FID (Feature Identifier) then becomes the subject of an RDF triple:

```
<osgb0000001234 rdf:type http://www.ordnancesurvey.co.uk/
ontology/v1/BuildingsandPlaces.owl/#Building>
```

the approach taken by D2RQ (Bizer and Seaborne 2004). The second option is that taken by Oracle (Lopez and Annamalai 2006), OWL Instance Store (Bechhofer, Horrocks, and Turi 2005) or RDF Triple Stores (Harris and Gibbins 2003), where the RDF triples* or OWL individuals are stored directly in a relational database. Finally, there is the choice of using Semantic Web services (Tanasescu et al. 2006; Cabral et al. 2006).

---

* A subject-predicate-object triple is the basic unit of an RDF data item, e.g. "Glen" is-a "Human".

D2RQ (Database to RDF Query) is a declarative mapping language for describing the relations between an ontology and a relational data model. Database content is mapped to RDF by a customizable mapping that specifies how the contents of a table are converted to subject-predicate-object triples. For example, the primary key field of a table row is identified as the subject of a triple, the column name is the predicate, and the column value is used as the triple's object. The SPARQL (Prud'hommeaux and Seaborne 2006) interface to the D2R Server (Bizer and Cyganiak 2006) enables applications to query the database using the SPARQL query language over the SPARQL protocol. The server takes requests from the Web and rewrites them via a D2RQ mapping into SQL queries against a relational database. The results are then reformatted for presentation back to the user as RDF triples, the result of the SPARQL query. This on-the-fly translation allows clients to access the content of large databases without having to replicate them in RDF. This method has the advantage that existing relational data do not have to be converted into RDF explicitly; however, one limitation is that SPARQL does not recognize spatial queries. A user-defined SPARQL function could be written to deal with a spatial query, but there would be no standard implementation, and the knowledge of how the spatial query had been defined would be buried in the SPARQL function. This solution goes against the principle of explicitness that is a major advantage of semantic technology. The same problem of burying knowledge arises with the D2RQ mapping file: although it can be a one-to-one mapping from an ontology class to a database table, and from an ontology property to a database column name, more complex mappings are allowed. For example, an ontology class can be mapped on to a subset of rows in the database table by generating a SQL WHERE clause. Specifying this information requires knowledge of the mapping file format and SQL, which takes it out of the hands of the domain expert, allowing content errors to more easily creep in. Tools like Topbraid (Topbraid Composer, http://www.topbraidcomposer. org) automatically generate the mapping file and "data ontology" from the database schema, but for complex databases like Ordnance Survey's, particularly ones with many spatially related system tables, these are generally unreadable. We are more encouraged by the approach of Perez de Laborda and Conrad (2005) that, unlike D2RQ, introduces the principle of specifying the mapping to the database within the ontology itself.

An alternative to creating a virtual RDF graph is to store RDF triples or OWL individuals directly in a relational database. Although the OWL Instance Store (Bechhofer, Horrocks and Turi 2004) or RDF Triple Store (such as the one described in Harris and Gibbins 2003) do not enable any spatial processing, Oracle's offering (Lopez and Annamalai 2006) allows the standard Oracle spatial SQL to be combined with their RDF_MATCH PL/SQL function, to perform a spatial query on data retrieved from an RDF table. The database contains geometry held within Oracle Spatial tables and attribute data are stored as RDF triples using Oracle's RDF support. At present the spatial component can only be executed after the RDF filtering. Our hypothesis is that in many cases it would be more efficient to perform the spatial query first to minimize the size of the RDF graph and hence make filtering more efficient. We would also like to see more reasoning capability and ontology expressiveness than just RDFS—although more is coming with Oracle 11g, which will allow a subset of OWL reasoning. The drawback of storing RDF directly in a database is twofold. First, it will be difficult to convince any company to convert all their relational data to RDF, and, second, RDF is likely to introduce significant performance degradation, due to the need to join the RDF table with the ordinary relational (spatial) data. It may be that RDF has most use as a data transfer format or to allow semantic querying, rather than sacrificing the tried and tested relational model for the RDF graph model.

It is also possible to coordinate Semantic Web services to perform spatial queries on multiple data sets at a semantic level. E Merges (Tanasescu et al. 2006) implements specific reasoning and data operations for an emergency planning application using Common Lisp, with the Internet Reasoning Service IRS-III (Cabral et al. 2006) (a platform that describes, publishes and executes Semantic Web services) written in the Web Service Modelling Ontology (http://www.wsmo.org/). There is a lifting and lowering module within IRS-III that "lifts" information from the XML output data of the Web service to create instances of the relevant ontologies, and a lowering function that creates XML data inputs to the Web services from ontology instances. Mediator descriptions provide mapping rules that align ontologies with data and select which services should be used to carry out the reasoning. While promising, this approach is quite inflexible, as the database can only be "seen" through the services, so the user can only extract from the database information that is supported by the services. There is also the question of how such a service-based system would scale when very large volumes of data are requested.

6.2.4 DATA ONTOLOGIES

We will now outline our approach to extracting data from a database in a way that can be understood semantically via a domain ontology: using ontologies to explain how to map the SPARQL query onto an SQL query, and hence retrieve the relevant data. Unlike D2RQ (Bizer and Seaborne 2004), we want to avoid any knowledge being hidden in a database-to-ontology mapping file, or the Java code that creates the SQL from the ontology, or, as presented in Uitermark et al. (1999), having a separate set of mapping or "abstraction" rules defined in Prolog. Therefore, we have developed an ontology describing a generic database, including concepts and relationships common to all databases such as Table, Column, Primary Key and Database Connection. This is similar to the Relational.owl ontology of Perez de Laborda and Conrad (2005), but because it avoids the use of classes as values, it can be represented in OWL-DL rather than OWL Full. This is important as it allows us to use one of the off-the-shelf inference engines for tractable reasoning.

This database ontology is imported by our data ontology, which defines the parameters of the specific database where our topographic data are stored. This includes information about how to connect to the database instance, along with details of how the data in the database can be understood in terms of the Buildings and Places domain ontology. A simplified example of this mapping is shown in figure 6.3. The concept "DB Building" in the data ontology is a subclass of the domain ontology "Building" and will be instantiated by data from the database, by using information in the data ontology to construct SQL queries to retrieve the

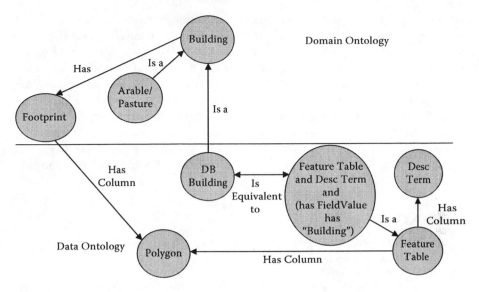

FIGURE 6.3 Simplified example of the relationships between a data and domain ontology.

relevant data from the TopographicArea Table. The necessary and sufficient conditions for "DB Building" (i.e. the bubble that is equivalent to "DB Building") can be converted to an SQL query to retrieve data to instantiate the domain ontology's Building class, while the necessary conditions for "DB Building" (i.e. the bubble that is a subclass of "DB Building") provides the information for how to construct an SQL query to retrieve data to instantiate properties of the Building, such as the Building's Footprint in this case.

6.2.5 AN EXPERIMENTAL RDF PRODUCT

By providing an OWL ontology for our existing products we are making the integration process easier, but the exchanged information itself will still be represented in a form that is specific to the GI community: GML (and internally everything is maintained in an object-relational database). However, if the information is instead represented in RDF format, then it can be more widely used. This is because RDF specifies explicit relationships between the data items, encoding some semantic meaning, rather than burying this information within the database schema and application code. We are therefore developing an experimental gazetteer described by an ontology and with the instance data represented as RDF.

6.3 CONCLUSIONS

Geographic information plays an important role in the overall information economy as an aid to the integration of nonspatial information. If the primary aim of a spatial information infrastructure is to serve the GI community alone, it could seriously impede the ability of GI to be adopted, manipulated and integrated by the wider information economy, condemning the GI community to a specialist backwater. We suggest that a spatial information infrastructure implemented using Semantic Web technologies will be able to operate across community boundaries, ensuring GI is adopted by the widest possible audience. However, the Semantic Web and its associated technologies are still very immature. Building ontologies is difficult and the technology required to link ontologies to databases is in its infancy, even before issues such as the support of spatial databases is considered. Ordnance Survey, as Great Britain's national mapping agency, has recognized the significance of Semantic Web technologies as a means to maximize the ability of our users to integrate GI into their business processes. We are therefore actively performing research into this area along with developing some initial topographic ontologies and experimental prototype products.

REFERENCES

Bechhofer, S., I. Horrocks, and D. Turi. 2005. The OWL Instance Store: System Description. In *Proceedings CADE-20, 20th International Conference on Automated Deduction*, Lecture Notes in Computer Science, Vol. 3632. Berlin: Springer-Verlag.

Berners-Lee, T., J. Hendler, and O. Lassila. 2001. The Semantic Web. *Scientific American*, May,

Bizer, C., and R. Cyganiak. 2006. D2R Server: Publishing Relational Data on the Semantic Web. Presented at the International Semantic Web Conference, Athens, GA, November 2006.

Bizer, C., and A. Seaborne. 2004. D2RQ: Treating Non-RDF Databases as Virtual RDF Graphs. Presented at 3rd International Semantic Web Conference (ISWC 2004), Hiroshima, Japan, November.

Cabral, L., J. Domingue, S. Galizia, A. Gugliotta, B. Norton, V. Tanasescu, and C. Pedrinaci. 2006. IRS-III: A Broker for Semantic Web Services Based Applications. Presented at 5th International Semantic Web Conference (ISWC 2006), Athens, GA, November 2006.

Doan, A., J. Madhavan, P. Domingos, and A. Halevy. 2002. Learning to Map between Ontologies on the Semantic Web. Presented at the Eleventh International WWW Conference, Hawaii, May 7–11, 2002.

Ehrig, M., and Y. Sure. 2004. Ontology Mapping: An Integrated Approach. In *Proceedings of the First European Semantic Web Symposium*, ed. Christoph Bussler, John Davis, Dieter Fensel, and Rudi Studer, 76–91. Lecture Notes in Computer Science, no. 3053. Berlin: Springer Verlag.

ISO/TC 211/WG 4/PT 19136 Geographic Information—Geography Markup Language (GML). Geneva: International Organization for Standardization.

Harris, S., and N. Gibbins. 2003. 3store: Efficient Bulk RDF Storage. Presented at 1st International Workshop on Practical and Scalable Semantic Systems, Sanibel Island, FL, October 20, 2003.

Hart, G., and C. Dolbear. 2007. What's So Special about Spatial? In *The Geospatial Web: How Geobrowsers, Social Software and the Web 2.0 Are Shaping the Network Society*, ed. Arno Scharl and Klaus Tochtermann. Advanced Information and Knowledge Processing Series. London: Springer.

Hart, G., C. Dolbear, and J. Goodwin. 2007. Lege Feliciter: Using Structured English to Represent a Topographic Hydrology Ontology. Presented at 3rd OWL Experiences and Directions Workshop 2007, Innsbruck, Austria, June 6–7, 2007.

Lopez, X., and M. Annamalai. 2006. Developing Semantic Web Applications Using the Oracle Database 10g RDF Data Model. Presented at Oracle Open World 2006, http://www.oracle.com/technology/tech/semantic_technologies/pdf/oow2006_ semantics_061128.pdf.

Manola, F., and E. Miller. 2004. RDF Primer W3C Recommendation February 10, 2004. http://www.w3.org/TR/rdf-primer/.

Open GIS® Geography Markup Language (GML). Encoding Standard, Open Geospatial Consortium. http://www.opengeospatial.org/standards/gml.

OWL Web Ontology Language Overview W3C Recommendation, D. L. McGuinness and F. van Harmelen, eds. 10 February 2004 http://www.w3.org/TR/owl-features/.

Randell, D. A., Z. Cui, and A. G. Cohn. 1992. A Spatial Logic Based on Regions and Connections. In *Proceedings 3rd International Conference on Knowledge Representation and Reasoning,* eds. B. Nebel, C. Rich, and W. Swartout, 165–176. San Mateo, CA: Morgan Kaufman.

Perez de Laborda, C., and S. Conrad. 2005. Relational.OWL: A Data and Schema Representation Format Based on OWL. Presented at 2nd Asia-Pacific Conference on Conceptual Modelling, Newcastle, Australia. *Conferences in Research and Practice in Information Technology,* Vol. 43, ed. Seven Hartmann and Markus Stumptner.

Perez de Laborda, C., and S. Conrad. 2006. Database to Semantic Web Mapping Using RDF Query Languages. Presented at 25th International Conference on Conceptual Modelling, Tucson Arizona, November 2006.

Prud'hommeaux, E., and A. Seaborne. 2006. SPARQL Query Language for RDF W3C. Working Draft. http://www.w3.org/TR/rdf-sparql-query/ 4 October 2006.

Runciman, B. 2006. Interview with Tim Berners-Lee. *ITNOW,* British Computer Society, March 2006.

Shariff, A. R., M. J. Egenhofer, and D. M. Mark. 1998. Natural Language Spatial Relations between Linear and Areal Objects: The Topology and Metric of English Language Terms. *International Journal of Geographical Information Science* 12 (3):215–246.

Tanasescu, V., A. Gugliotta, J. Domingue, L. Gutiérrez Villarías, R. Davies, M. Rowlatt, M. Richardson, and S. Stinčić. 2006. Spatial Integration of Semantic Web Services: the e-Merges Approach. Presented at Terra Cognita: Directions to the Geospatial Semantic Web. Athens, GA, November 2006.

Uitermark, H., P. van Oosterom, N. Mars, and M. Molenaar. 1999. Ontology Based Geographic Data Set Integration. In *Proceedings of the International Workshop on Spatio-Temporal Database Management STDBM 99,* ed. M. H. Bohlen, C. S. Jensen, and M. O. Scholl, 60–78. Lecture Notes in Computer Science, no. 1678. Berlin: Springer.

Uitermark, H., P. van Oosterom, N. Mars, and M. Molenaar. 2005. Ontology Based Integration of Topographic Data Sets. *International Journal of Applied Earth Observation and Geoinformation* 7:97–106.

7 Using Formal Semantics for Services within the Spatial Information Infrastructure

Rob Lemmens

CONTENTS

There is an increasing need for organizations to integrate and reuse geo-information and geoservices from within and outside the organization. These activities are typically performed in the context of the spatial information infrastructure (SII). The process of integrating services is commonly referred to as service chaining. This requires that services can be easily found, and that they are executable and interoperable. Interoperability means that the services "understand" each other's messages. A major impediment is formed by the semantic heterogeneity (the differences in meaning) of geo-information and of the functionality of geoservices.

Making services semantically interoperable is an important prerequisite for information sharing in today's networked society. This involves services that rely on different knowledge domains, one of which is the geo-information domain.

Within this context, this chapter provides solutions for the computer-aided integration of distributed heterogeneous geo-information and geoservices, based on their semantics (the meaning of their content).

Geo-information distinguishes from other information by its spatial relevance. Geoservices often have to deal integrally with multiple representations of features in a spatial, temporal, and thematic dimension. Geoservices are also implicitly connected by the geographic location of the features they process. This has implications for the interoperability of geoservices. For example, the validity of a service (e.g., a route planner) may be bound to a specific geographic area, which could imply it cannot be used in combination with services involving another validity area. On the contrary, services that seem to be incompatible due to differences in feature representation (e.g., geometry, coordinate reference system) may turn out to be useful in combination, because they contain information on the same locations.

In the geo-information domain, an ontological concept is typically constructed by the basic notion of a feature type and its properties (e.g., the geometric object that represents it). These constructs form the common ground (called the semantic interoperability framework) to which semantic descriptions of interoperable services should refer. These descriptions are expressed with ontological constructs. For the purpose of service discovery, we distinguish between requesting service descriptions and advertised service descriptions. The proposed semantic framework SIFGEO allows for descriptions with different detail and makes relaxed queries possible. The provision of concepts at the level of geographic features (e.g., "Building" [as real-world concept] and "Point" [as geographic representation]) is considered a minimum requirement for a semantic interoperability framework for geoservices. As a backbone, a basic ontology with general constructs is made available, based on the ISO General Feature Model (ISO 19109).

In addition to information modeling constructs, we need to model the characteristics of service functionality to support the exchange of explicit service capabilities. For this purpose, services are described by the operations that they make available. A model implementation is provided based on the Web Ontology Language (OWL) (McGuinness and van Harmelen 2004). The following basic elements are considered to be the essentials that can be modeled semantically and of which at least one should make part of a geo-operation characterization:

- Classification of geo-operation functionality
- Description of operation input and output parameter types
- Description of geodata that are tightly coupled to the service
- Description of the control flow in (virtual) composite operations

The above elements of geo-operation descriptions should be referenced to formally defined elements in what we call "semantic interoperability frameworks," which are based on ontology technology. The remainder of this chapter describes what is meant by semantic interoperability frameworks (section 7.1) and how they can be applied for modeling geo-information and geo-operations. A framework designed by the author, called SIFGEO, is presented in section 7.2. Ontological design issues are discussed in section 7.3, in which an ontology for geo-operations named OPERA is proposed as part of SIFGEO. An example usage of the framework is provided in section 7.4 in the form of a small case study. The chapter ends with a discussion on the deployment of the framework (section 7.5).

7.1   SEMANTIC INTEROPERABILITY FRAMEWORKS

7.1.1   DEFINITIONS

Ontologies do not come as a single solution to a demand for information integration. They are typically embedded in a framework and an infrastructure. The definitions below are adaptations of the notions of framework and infrastructure that are used in Stuckenschmidt and van Harmelen (2005) to characterize a specific information sharing approach. In this chapter, they are defined as follows:

A semantic interoperability framework is the combination of ontologies, their relationships, and methods for ontology-based descriptions of information sources (data sets, services, etc.). The framework serves the semantic interoperability between information sources.

A semantic infrastructure is defined as follows: a semantic infrastructure comprises a semantic interoperability framework and the tools to maintain and use the framework as well as the (meta-)information sources that are produced with these tools.

Important aspects of building, maintaining, and using a semantic infrastructure are

1. Ontology creation and access
2. Ontology integration
3. Ontology-based description of information sources (annotation)
4. Reasoning-based information retrieval, semantic translations, and information integration/fusion
5. Creation and use of ontology meta-information (information about ontologies)

7.1.2   FORMAL MODELING OF OPERATIONS AND THEIR CONTROL FLOW IN OWL-S

OWL-Services (Martin et al. 2004), or OWL-S in short, is an upper-ontology based on OWL that models the characteristics of Web services and which can be used to create semantically enriched Web service descriptions.

OWL-S provides three modeling constructs at the top level, that is, the service profile (what the service does), the service grounding (how the service can be accessed), and the service model (how to use the service in terms of semantic content, including its workflow). These three basic models are depicted in figure 7.1.

OWL-S provides classes that can be instantiated by a service provider to create specific service descriptions. This implies that such descriptions are expressed as OWL individuals in all three OWL-S subontologies. Because OWL-S is an upper ontology, it obviously does not provide domain ontologies. These have to be established by service communities themselves.

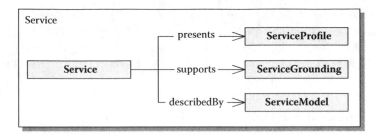

FIGURE 7.1 UML class diagram showing the overall service model of OWL-S. (From Martin, D. et al. 2004. OWL-S: Semantic Markup for Web Services. W3C member submission, expository document, World Wide Web Consortium. With permission.)

The process model of OWL-S (and its implementation in machine ontology) is based on principles that have been developed in previous work on process modeling, among others pi calculus, the Process Specification Language (PSL), and Golog (Martin et al. 2004). Figure 7.2 shows the basic structure of the OWL-S process ontology. Note that the service model class appears in both figures 7.1 and 7.2. A participant in OWL-S is a client or a server. OWL-S supports the paradigm of IOPE parameters (Input, Output, Precondition, and Effect). For preconditions and effects it

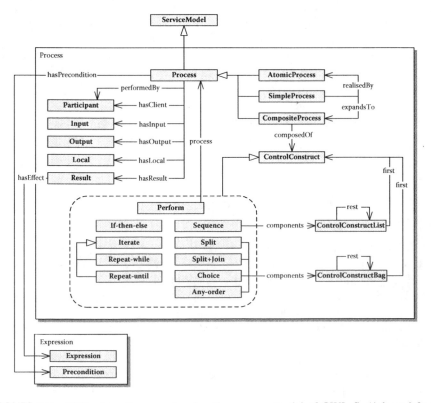

FIGURE 7.2 UML class diagram showing the process model of OWL-S. (Adapted from Martin et al. 2004; Barstow et al. 2004.)

uses local parameters and expressions, which can be declared in a specific language, such as SWRL (Semantic Web Rule Language) (Martin et al. 2004).

A process can be one of three types, that is, atomic, simple, or composite. An atomic process is defined in Martin et al. (2004) as a description of a service that expects one (possibly complex) message and returns one (possibly complex) message in response. Atomic processes are directly invocable; they have no subprocesses and execute in a single step, as far as the service requester is concerned. Composite processes can be decomposed into other composite or noncomposite processes. A simple process is an abstraction of an atomic or a composite process.

The way in which a composite process is constructed is depicted in the middle part of figure 7.2. A control construct can be any of nine control flow types. A composite process is constructed in the form of a tree with branches following either "first" or "rest." The "Perform" construct is used to instantiate the branches (i.e., all branches and leaves are expressed as OWL individuals).

7.2 SEMANTIC INTEROPERABILITY FRAMEWORK FOR GEOSERVICES (SIFGEO)

This section describes the basic elements of a semantic interoperability framework for geoservices named SIFGEO, based on the principles of semantic interoperability frameworks, discussed in section 7.1. It is based on Lemmens (2006) and describes the framework of formal ontologies and the way they support the characterization of geo-information and geo-operations for the purpose of machine reasoning. The formal semantics of information and operation concepts is expressed in Description Logic axioms and implemented as machine ontologies in OWL.

7.2.1 GEOSEMANTIC MODELING AND SPATIAL RELEVANCE

We should ask ourselves whether the semantic modeling of geoservices is essentially different from semantic modeling in general. The essential question is whether the information retrieval process has a spatial relevance. In fact, it has with respect to several aspects. Although the research work that forms the basis for this chapter does not embark upon spatial reasoning, the most prominent aspects and consequences for semantic modeling are highlighted below (see also Lemmens et al. 2006):

- Geo-information is meant to exhibit spatial relationships between features. These relationships are used for spatial analysis in GIS by computations on topology and metrics of geometries. Currently, ontology languages such as OWL do not contain specific constructs that model spatial relationships. As a workaround, common practice is to specify roles with a spatial connotation, such as "Touch," "Overlap," and "North." Examples can be found in Arpinar et al. (2006). Another alternative is to outsource the spatial analysis to conventional computational solutions.
- Geo-information is multidimensional. The integrated spatiotemporal and thematic aspects of geo-information contribute to its importance in a wide variety of application domains. At the same time, the interrelation

of aspects introduces complexity of reasoning over multiple aspects when geo-information is involved. Geoservices make use of operations that may act on all dimensions of geo-information. For a proper understanding of the functionality of those operations, it is important to identify these dimensions in an operation.

- Geo-information is characterized by multiple representations. Typical for geographic information (and geoservices) is that it is very common for a geographic phenomenon to take many different feature representations in multiple or even a single geodata set. Representations may differ in terms of spatial, temporal, and thematic attributes. Such attributes may also vary along different levels of generalization and aggregation. Reasoning about geoservices has to take into account these aspects, specifically with respect to the meaning of its input and output parameters.
- Geoservices often depend on tightly coupled geodata. Tightly coupled data determine the validity area of a service and may contain (part of) the semantics of the service's input and output parameters.

The above aspects have been taken into account in the implementation of the proposed semantic interoperability framework and determine the essence of the reasoning about geo-information and services.

7.2.2   SEMANTIC FRAMEWORK OVERVIEW

This section gives an overview of SIFGEO and its ontologies that we apply for the modeling of geo-information and geo-operations as contained in geoservices. The scope of the ontology framework covers both geo-information and geoservices. The ontology framework can be used for geoservice discovery and service chaining, by means of ontology queries. Typical queries that can be posed to this ontology framework are

- Find all operations that match a set of input and/or output parameter types.
- Find all operations that fit an existing service chain with respect to their input and/or output parameter types.
- Find all operations that are composed of operations that instantiate a given set of operation types.
- Find all information/service concepts that are sub- or superclasses of a given concept.
- Find all data sets that contain a specific feature type (e.g., *Building*).

7.2.3   FRAMEWORK ONTOLOGIES

At the basis of SIFGEO, there are three types of formal ontologies: feature concept, feature symbol, and geo-operation ontology. They are introduced briefly below:

- A feature concept ontology formally defines the conceptualizations of real-world phenomena and the relationships between them. Examples of elements in the feature concept ontology are *Building* and *ConstructionMaterial*. The elements of a feature concept ontology make part of an application schema (the conceptual schema for data required by one or more applications (ISO/ TC211 2001). An element in the feature concept ontology is defined by the relationships with other elements in the feature concept ontology (for example, *Building* is a subclass of *Construction*) and by the relationships with elements contained in the feature symbol ontology (for example, *Building* is represented in a particular application schema by a point). The latter relationship is represented by the *hasGeometry* relationship for the specific example of *Building*. Generally speaking, the feature concept ontology "uses" the classes of the feature symbol ontology for its definitions.

- A feature symbol ontology formally defines the elements that make up a feature at a "symbol" level and the relationships between them. The term "symbol" does not necessarily refer to a visual symbol, but rather to a semantic symbol. Examples of these elements are *GF_FeatureType* and *Point*. A feature symbol ontology may also contain instances at the data level. For example, an instance of the class *Point* is an actual point with coordinates.

- A geo-operation ontology formally defines types of operations in terms of their behavior. Each type is characterized by the behavior of one out of a set of well-known atomic GIS operations and their typical input and output parameters. GIS operations may also be tightly coupled to a data set, for example, a road data set in the case of a shortest path operation. The input and output parameter types are described by referring to elements from the feature symbol ontology. For example, these elements may indicate that a service needs thematic attributes to function properly (and, for example, does not need spatial attributes). Further, for composite operations, the ontology contains control flow elements, such as Sequence and Choice.

Each of the above three ontology types may be materialized by a specific ontology or a combination of ontologies. A specific feature symbol ontology may reflect how a specific set of services of a software manufacturer handles its geographic feature representations. Another feature symbol ontology may implement standardized geographic feature representations, such as defined in ISO or OGC specifications. Similarly, geo-operation ontologies may be manufacturer dependent (proprietary) or contain standardized elements. With respect to feature concept ontologies, we do not differentiate between proprietary and standardized ones, but rather between generic ontologies and domain-specific ones. The application domain that a semantic framework has to serve requires the feature concept ontology to include a particular scope. For example, a semantic framework that serves a traveling domain requires a feature concept ontology that defines traveling concepts. A feature concept ontology is typically built by experts within information communities and tends to have a limited scope.

7.2.4 RELATIONS WITH ISO SPECIFICATIONS

SIFGEO, as introduced in the earlier sections in this chapter, is based on a number of ISO standards for geo-information. They are introduced briefly here. The ISO 19131 standard for data product specifications provides guidelines for creating data product specifications (such as those of national mapping agencies) in terms of other existing ISO specifications. Application schemas and feature catalogues are given a central role in representing the content of data. An application schema defines the data structure and the data content in accordance with ISO standard 19109 (General Feature Model). The application schema implements a feature catalogue, which provides the semantics of feature types, their attributes, attribute values, feature behavior, and relationships between features. Specifications for creating a feature catalogue are described in ISO standard 19110 (Methodology for Feature Cataloguing). Both the ISO models for application schema and the feature catalogue draw upon the rules given in the ISO General Feature Model. The new Dutch geo-information model NEN3610 (version 2) is following the above ISO standards. Apart from the application schema and feature catalogue, the ISO 19131 (Data Product Specifications) further specifies the inclusion of other elements such as data delivery parameters and data quality parameters.

A feature symbol ontology as introduced earlier can be considered as a representation of a feature model and implements the ISO General Feature Model. A feature concept ontology represents a feature catalogue. This may be a feature catalogue conforming to the ISO 19110 (Methodology for Feature Cataloguing) standard. The ISO 19119 (Services) standard is relevant with respect to operation metadata and operation type classification.

7.3   GEO-OPERATION CHARACTERIZATIONS: OPERA

This section describes the starting points for the design of an ontology of geo-operations, called OPERA, as part of the SIFGEO framework. From a design perspective, one would like such an ontology to (1) have a hierarchical structure (for ease of human understanding), (2) have nonoverlapping classes as much as possible, (3) include the most important geo-operations, and (4) be extensible. In the past, several attempts have been made to create a classification of geo-operations. The most relevant ones are stemming from research by Albrecht (1995, 1998) and Chrisman (1999, 2002), from standardization efforts, such as ISO 19119 (Services), and from software models (Goodchild 2001). Most of them are, however, informal and lack sufficient semantics for classifying individuals. A major problem was caused by the lack of sufficient mechanisms to support multiple inheritance. With modern ontology languages such as OWL, we can now overcome this problem.

OPERA is based on the principles of OWL-S and particularly implements its process model. However, specific to OPERA is its classification of geo-operations and its descriptions of specific geo-operation parameter types and tightly coupled data, all of which are not part of the OWL-S ontology. The main strategy in classifying geo-operations is to consider the elements of the feature concept ontology and the feature symbol ontology to be the basic representatives for the input and output parameter types of the operation classes.

7.3.1   ATOMIC GEO-OPERATIONS

The design of a geo-operation ontology in this chapter (based on the research described in the Ph.D. thesis of the author) is based on a distinction between atomic geo-operations and composites of them. The definition of an atomic geo-operation, as provided below, follows the definition of an atomic process in OWL-S. In OWL-S, an atomic process is defined as a directly invocable process that executes in a single step. It takes an input message and returns an output message. These messages can take as many formal input and output parameters as required.

An atomic geo-operation may include functionality also performed by another atomic operation. This does not make it a composite. For example, consider an overlay operation that is performed on a feature collection of which the topological relationships between its features are not made explicit. As part of the operation it must first calculate the intersections between the features and then combine the thematic attributes. This (sub)operation could be performed by the atomic operation *MakeTopologyRelationshipsExplicit*. The overlay operation is still considered to be atomic. Further, an atomic geo-service is considered to be a service that makes available an atomic geo-operation and is not composed of other services.

A distinction has to be made between operation implementations and operation descriptions. An operation implementation is the invocable software artefact (e.g., a software component) that carries out the operation. A description is considered to be represented by (1) a classification of the service as an operation class in the geo-operation ontology and/or (2) a workflow with its suboperations, each of which is also classified as a class in the geo-operation ontology.

7.3.2 Ontology Design—OPERA-R and OPERA-D

Our geo-operation ontology (OPERA) is developed with two sets of concepts:

- A reference geo-operation ontology (OPERA-R), containing atomic geo-operation types that act as building blocks for all other geo-operation types.
- Derived geo-operation ontologies (OPERA-D), which may contain
  - atomic geo-operation types, each of which is defined in terms of an operation type, existing in the reference ontology (OPERA-R).
  - composite geo-operation types, each of which is defined in terms of a workflow and its component operation types, being geo-operation reference types or derived types.

An overview of the main structure of OPERA is given in figure 7.3. At the root is the class *GeoOperationType* with two specializations, OPERA-R and OPERA-D. OPERA-R has five direct subclasses that involve, respectively, human interaction operations, feature modeling operations, feature operations, operations on services, and meta-information operations.

OPERA-R describes atomic geo-operation types. Each type can be instantiated by an atomic geo-operation instance. A more detailed description of the informal semantics of the classes of OPERA-R is provided in Lemmens (2006).

The class *FeatureProcessingOperation* has been selected to be specified in more detail as it forms a broad basis for geo-information processing and analysis activities in practice. In contrast to feature modeling operations, feature processing operations are not meant to add feature types to an application schema, although they may

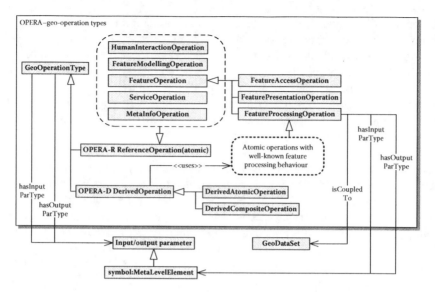

FIGURE 7.3   Generalized structure of the OPERA geo-operation ontology.

create feature types and feature properties as placeholders for feature instances and their attributes. For example, an operation of type *Buffer* may have an output parameter, named *BufferZone*, of type *GF_FeatureType* with *GF SpatialAttributeType* as a property type. An instance of this operation (e.g., *MyBufferOperation*) has an output parameter of type *FeatureType* with an attribute of type *Polygon*. At the operation invocation level, the output of this operation may be an actual buffer instance, represented by an actual polygon with coordinates.

The input and output parameter types of a feature processing operation are expressed in terms of elements of the feature symbol ontology, described in section 7.2.3. A *FeatureProcessingOperation* may be coupled to a data set, which may constrain the validity of usage of that operation. For example, a specific route planner operation may operate within the boundaries of a national data set. Feature processing operations are discussed in more detail in the next section.

7.3.3   OPERA-R: FEATURE PROCESSING OPERATIONS

This section describes types of feature processing operations as part of SIFGEO. The feature processing operations in this framework apply the OWL-S process model. This implies the distinction between atomic and composite operations. In OWL-S, an atomic feature processing operation does not maintain state, but a composite operation does.

Atomic feature processing operations and composite feature processing operations borrow the semantics from, respectively, OWL-S atomic and composite processes.

7.3.4   CLASSIFICATION OF OPERATIONS

An overview of feature processing operations is provided in figure 7.4. Each operation class is given a short name for easy reference, such as *Change-CRS* (a class of operations that change the coordinate reference system of a coordinate set). The operations are grouped in sections. This classification is based principally on the kind of feature properties that they act upon or return. The classification of GIS operations by Chrisman (2002) follows for the major part the same principle and is used in particular in this chapter for the description of attribute operations, overlay, and distance-based operations. Other parts of that classification are not directly followed, such as his concept of transformation. Chrisman defines a transformation as an operation that changes a measurement framework (a scheme with measurement rules). This mechanism is rather complex and seems to lead to an ambiguous classification of operations. The concept of transformation that is used in the research that forms a basis to this chapter classifies geometric transformations as operations that change the coordinates of positions.

In some cases, an operation type may be classified under more than one section. For example, the *GridSlope* operation type is classified under both categories *GridFilter* and *CalculateSlope* (both subclasses of the Neighborhood operation type; only the latter is displayed in figure 7.4), which implies that these two categories are not mutually exclusive.

7.3.5 OPERATION TYPE DESCRIPTIONS

Each operation type as depicted in figure 7.4 is characterized by its functional description, its input and output parameters, and its tightly coupled data. The input parameters attached to an operation type by the *hasInputPar* are the formal parameters minimally needed for that operation type. The output parameters attached to an operation type by the *hasOutputPar* are the formal parameters that are minimally needed to contain the output of the operation type (an example will be given below for the operation "LocSpat"). Operation instances may make use of other parameters, but they are irrelevant for describing the operation type. With respect to parameter passing, a single formal parameter may be used to pass multiple values (e.g., instances of geometric objects) in an array. In addition to the generic classes provided in figure 7.4, specific descriptions can be created by describing subclasses of generic operation

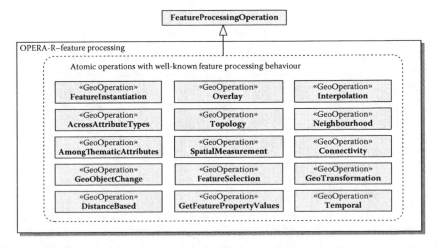

FIGURE 7.4   Feature processing operation classes in the OPERA-R ontology.

types. For example, a more specific description of a route optimization operation may specify the number of given points in a route. The input/output model assumed here follows the one of OWL-S, which supports stateless services (with input and output parameters) as well as stateful services (with, in addition, pre- and/or post-conditions). A feature processing operation may require a data set to comply with a precondition, that is, to be in a specific state. This may involve the features being processed, but also any other feature in the data set. For example, the entire data set must be defined in a specific coordinate system. A post-condition may be of a similar kind. The ontology also contains so-called support concepts. They define the general characteristics of operations such as operation-specific details, like a property named topological selection method that is (only) attached to the overlay operation.

An example of an operation definition that is used for instantiation and reasoning is given below for the *LocSpat* operation. The name *LocSpat* represents an operation type that reads a location attribute type (e.g., instantiated as an address type) and produces a spatial attribute type (e.g., instantiated as a geometric object type), which is typically found in a gazetteer:

```
opera:LocSpat ⊑

    opera:AcrossAttributeTypes ⊓
    (∃∀  opera:appliesToDataStrucType.(symbol:ObjectFeature  ⊔
symbol:GridCell)) ⊓
    (∃∀ opera:hasInputPar. (∃ opera:hasParType.symbol:GF Loca-
tionAttributeType)) ⊓
    (∃∀  opera:hasOutputPar.  (∃  opera:hasParType.symbol:GF _
SpatialAttributeType)) ⊓
    (≥1 opera:isCoupledToDataset))

with:

    ⊑        "is subclass of"
    ∃∀       conjunction of "there exists at least one" and "for all"
    ⊓        "intersected with"
    ⊔        "union of"
    ≥1       "has at least one"
    .        separator between role and role-filler
```

The above operation description is represented in Description Logic. In the above example, the *opera:hasInputPar* is conditioned by an existential quantification to state that the *LocSpat* operation needs at least one parameter of type *symbol:GF_LocationAttributeType*. The value restriction is used to make sure that any input parameter is of type *symbol:GF_LocationAttributeType* and nothing else. The cardinality restriction of the role *opera:isCoupledToDataset* indicates that the operation is coupled to at least one data set.

7.4   CASE STUDY

This section describes the implementation of a use case that involves a researcher named Jody who creates a service chain for the purpose of satellite image processing and a researcher named Jeff who reuses this service chain. The service chain serves to evaluate the extent of flooding in the confluence area of the Ganges and Jahmuna rivers in Bangladesh.

In the first phase, Jody has the role of service developer. For creating the services, she uses three SPOT images, representing, respectively, the periods during the dry season, a moderately severe flood, and a severe flood. In this phase, it is assumed that the processing services have been identified and that they are directly available. As Jody wants to describe her method and wants to make it available to others, she creates meta-information for each component service and for the composite service (a sequence). As each service contains only one operation, this is done by referencing each service to a class in the OPERA ontology. The result is an OWL-S document that describes the service chain semantically. The subsequent services are described below:

1. Jody uses a *band rationing service* to distinguish land-water boundaries. Dividing SPOT band 3 by band 1 is a known method that accomplishes this. This service can be classified as *opera:CrossCalculate* in the OPERA ontology.
2. A *slicing service* is used to classify the water and land pixels in each SPOT image. This service is a subclass of OPERA's *opera:Classify*.
3. With a *cross service*, the three land-water grid coverages are now combined into one, by creating a class for each combination of land and water attributes The cross service is classified as *opera:CrossCalculate* in OPERA.
4. Finally, Jody uses a service called *ConvertToClasses* to show the impact areas of two flooding periods. The service is a subclass of *opera:Group* With an annotation tool (in this case an OWL-S editor), she imports the OPERA ontology and enters the service chain definition below:

EvaluateFlood ≡ C

in which *C* is an ontological concept that describes the control flow pattern below (the first colon in each line means "is of type"):

```
<sequence>

        BandRationing : opera:CrossCalculate
        Slicing : opera:Classify
        Cross : opera:CrossConcatenate
        ConvertToClasses : opera:Group

</sequence>
```

7.5   DEPLOYMENT AND CONCLUSION

Geoservice discovery involves the identification of service advertisements that match a service request. Apart from behavioral aspects, matches are sought between requested and advertised input/output parameters. Matchmaking may involve service parameter

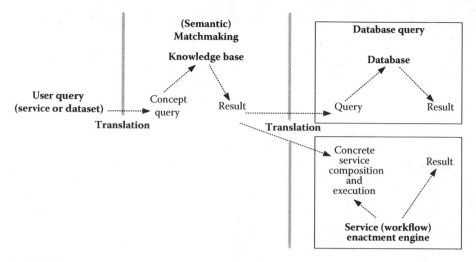

FIGURE 7.5 Semantic matchmaking provides conceptual links between information models and may form an intermediate between a user query and a database query or service execution. It does not query data directly in the database, nor does it invoke a service directly.

concepts as well as data set concepts. Further, matchmaking, as described here, is done in a semantic framework. The ontological concepts and individuals involved are abstract representatives of the actual service parameters and/or database entities. Despite the fact that ontologies are capable of also containing the concrete parameters and entities, they are not used for this purpose here. Figure 7.5 shows that there needs to be a translation between a user query and a query, posed to the knowledge base, and a translation of the result of the latter to either a database query or a service invocation.

As the proposed semantic interoperability framework, SIFGEO has been developed in an obviously conditioned prototype environment; its immediate deployment in a specific application requires the creation of a semantic infrastructure. In such an infrastructure, the presented feature symbol ontology and geo-operation ontology (OPERA) are immediately deployable.

In addition, an application ontology has to be built for the specific application at hand. This application ontology may refine OPERA by adding subclasses of operations that are specific to the application. It may also need to extend existing domain ontologies with respect to feature concepts. Existing domain ontologies may be found in ontology libraries, such as SchemaWeb (http://www.schemaweb.info). Once the proper ontologies are found, a complete set of mappings needs to be established between them. In this form, the framework can function as a basis for semantic service descriptions.

Application fields that are thought to benefit from the developed theory in the short term are

1. Harmonization of super- and sub-models; creation of ontology-based meta-information and application of matchmaking.
2. Generalization of geographic features (such as in database/geographic map generalization). Semantics are needed for making the right decisions with respect to (re)moving and grouping of feature instances.

3. Data set and service discovery as part of catalogue mechanisms in SIIs at all organizational levels.
4. Semantic translation of information (from one application domain to another).
5. Ontology presentation to users of a specific geographic data model. This requires a comprehensive visualization tool.
6. The matching of user profiles with service advertisements, for example, to create context awareness in Location Based Services (LBS) applications.
7. Quality assessment and improvement of metadata for data sets and services.

Reasoners can check the consistency and completeness of ontology-based information source descriptions, by integrally inferring the ontological relationships that exist between them.

The offered solution is flexible and extensible. With respect to flexibility, the framework allows the use of incomplete service descriptions. With respect to extensibility, the framework allows service descriptions that can be extended with new concepts. Moreover, existing application domains can be linked through ontology mappings. In the process of service chaining, four steps have been identified: discovery, abstract composition, concrete composition, and execution.

The link between the abstract and concrete composition of services is realized by annotation, which connects ontology elements with parameters of executable code.

The deployment of the approach requires key organizations such as OGC to develop and maintain domain-independent parts of a semantic interoperability framework and organizations with an SII mandate to manage its domain-dependent parts.

REFERENCES

Albrecht, J. 1995. Universelle GIS-Operationen. Ph.D. thesis, Fachbereich Sozial- und Kulturwissenschaften der Hochschule Vechta.

Albrecht, J. 1998. Universal Analytical GIS Operations: A Task-Oriented Systematization of Data Structure-Independent GIS Functionality. In *Geographic Information Research: Transatlantic Perspectives,* ed. M. Craglia and H. Onsrud, 577–591. London: Taylor & Francis.

Barstow, A., J. H. M. Skall, J. Pollock, D. Martin, V. Marcatte, D. L. McGuinness, H. Yoshida, and D. D. Roure. 2004. OWL-S Ontologies. World WideWeb Consortium Web site. Available at http://www.w3.org/Submission/2004/07/.

Chrisman, N. 1999. A Transformational Approach to GIS Operations. *International Journal of Geographical Information Science* 13 (7):671–637.

Chrisman, N. 2002. *Exploring Geographical Information Systems,* 2nd ed. New York: John Wiley & Sons.

Goodchild, M. F. 2001. Spatial Analysis and GIS. In *Proceedings Environmental Systems Research Institute (ESRI) User Conference Pre-Conference Seminar.*

ISO/TC211. 2001. *Final Draft International Standard 19101 Geographic Information—Reference Model.* Tech. Rep. N 1197, International Organization for Standardization, Technical Committee ISO/TC 211, Geographic Information/ Geomatics. Geneva: International Organization for Standardization.

Lemmens, R. L. G. 2006. Semantic Interoperability of Distributed Geo-Services. Ph.D. thesis, International Institute for Geo-Information Science and Earth Observation (ITC), Enschede, The Netherlands.

Lemmens, R., C. Granell, A. Wytzisk, R. de By, M. Gould, and P. van Oosterom. 2006. Integrating Semantic and Syntactic Descriptions to Chain Geographic Services. *IEEE Internet Computing* 10 (5):42–52.

Martin, D., M. Burstein, J. Hobbs, O. Lassila, D. McDermott, S. McIlraith, S. Narayanan, M. Paolucci, B. Parsia, T. Payne, E. Sirin, N. Srinivasan, and K. Sycara. 2004. OWL-S: Semantic Markup for Web Services. W3C Member Submission, expository document, World Wide Web Consortium. Available at http://www.w3.org/Submission/2004/07/.

McGuinness, D. L., and F. van Harmelen. 2004. *OWL Web Ontology Language Overview.* W3C Recommendation, World Wide Web Consortium. Available at http://www.w3.org/TR/owl-features/. Accessed December 2005.

Stuckenschmidt, H., and F. van Harmelen. 2005. *Information Sharing on the Semantic Web.* New York: Springer.

8 Geosemantic Web Standards for the Spatial Information Infrastructure
Nice to Have or Hopeless Without?

Joshua Lieberman and Chris Goad

CONTENTS

Consideration of geosemantic Web standards for a spatial information infrastructure (SII) raises a series of questions: What is the nature of spatial information infrastructure? What role do standards play in promoting interoperability? What is the balance between imposition of uniform information standards and semantic mediation between disparate information standards? This chapter examines the case that reasonable expectations for effective SIIs do require intrinsic support for spatial information interoperability. Geosemantic (i.e. geographically grounded spatial semantic) interoperability is an especially unique aspect of SII requirements; it is shown to depend upon many more subtleties of meaning and levels of interoperability than just shared vocabularies. Particular geosemantic interoperability "pain points" in SII are illustrated by simple examples of cognitive and pragmatic heterogeneity in spatial information sharing. Although geosemantic technologies show some promise of reducing the extent of this reliance, their present state of development suggests that SII will need to rely, at least in the near term, upon widespread adoption of standard geosemantic foundation ontologies. Categories of geosemantic ontologies that are the most likely candidates for initial geosemantic Web standards include features,

119

feature types, feature data sets, spatial relationships, place names (toponyms), coordinate reference/spatial index systems and spatial services.

8.1   INTRODUCTION

Any information infrastructure that is built or sustained by way of more than one vendor or technology relies critically on standards. Standard formats and protocols enable disparate components to function at some level as a connected whole. The question is, at which level and for what function? The term "infrastructure" usually implies "lower" or "underlying" structure, a common support for higher and more application specific functionality. Somewhere among the higher levels of functionality and interoperability are usually included the exchange of meaning, the harmonization of theories and the agreement on intentions. Among the most specialized issues dealt with at these levels are ones concerning location, geography and other spatiotemporal characteristics of entities in the physical world.

Within a restrictive definition of infrastructure, standards for SIIs should be restricted to those governing management and transport of generic information sets, such as Dublin Core, TCP/IP, HTTP and perhaps ebXML. Geographic information and semantic standards would become relevant not for the (S)II itself, but only within specialized applications built on top of the (S)II. Such a systematic decomposition is suggested by many who question just what is "special about spatial" (Hart 2006). It is not certain, however, that a clear distinction of infrastructure exists within the realm of information systems that is analogous to that between a train track and a train wheel. Taking these circumstances as an opportunity, it is worth considering what our operational expectations are for (S)II and therefore what possible infrastructure distinction might best promote both interoperability and reuse of components in spatial information systems.

8.2   ROLE OF SPATIAL INFORMATION INFRASTRUCTURE

An appropriate decomposition for SII comes down to this question: what do we want from our infrastructure? Is it the ability to move large amounts of information long distances (*tracks and trains*) or is it the ability to move just the right information from a provider to a consumer even when they may neither know of nor fully understand each other (*optimal logistics*)? The answer is not obvious and points to an important dilemma. On the one hand, infrastructure by its nature is intended as a common capability that can be used by many parties for many applications; what constitutes the *right* information is usually dependent on the specific application rather than being common to all. On the other hand, a capable but indiscriminate information transport infrastructure can quickly take us from having insufficient information to being flooded by it. Is this a real problem? After all, generally the most thorough means of wading through and analyzing information is locally within one consistent, complete information base. Why not take advantage of ever increasing transport and storage capacity to do this?

There at least two problems with this. First, there are clearly combinatory issues with supporting an increasingly complex web of information relationships on top of any physically distributed transport infrastructure. This is especially true for an information infrastructure that is unable to do spatial optimization. Only so many whole-earth data sets can be passed around any network in near-real-time! Indeed, most spatial data sets are not intrinsically whole-Earth, but are either localized or specialized. They tend to be built and maintained by a local or specialized organization. Regional or global "roll-up" of up-to-date spatial data, whether physical or virtual, is both a common and a significant task.

Second, it is unclear that any volume of information will necessarily lead to greater understanding or utility of information if the basis for both relevance and comprehension is not shared between providers and consumers. For example, lacking a common geosemantic basis for CAD and GIS data, no volume of information can resolve either which distributed data sets cover the same features, or how providers/consumers of either type of data should interpret examples of the other type. This concept is expanded below. Either of the above problems imposes serious limitations on the benefits we can hope to receive from infrastructure design and investments under a restrictive and generic model.

8.3   SPATIAL INFORMATION INTEROPERABILITY: LAYERS AND PAIN POINTS

When we speak of commonality, accessibility and reuse of spatial information or spatially enabled computing resources, we are in large part speaking about spatial information interoperability. This is a general and somewhat ambiguous concept that is difficult to define formally. Interoperability itself is a developing area of scientific or applied philosophic inquiry. An informal view might hold that an interoperable information system is one that is resilient to changes in its subsystems, components or participants, for example, exchanges of audience, protocol, technology, application and so on. In particular, interoperability implies a resilience to heterogeneity, the vendor of a server being different from that of a client, or the community originating information being different from that consuming it.

8.3.1   INTEROPERABILITY LAYERS

That there can be many levels of interoperability is frequently expressed in the layered nature of communications and transport protocols. Each higher level adds a more detailed communication function and is dependent on the effectiveness of the layers below it to provide less detail but greater resilience to change. Figure 8.1 is a conceptual synthesis of such layering. Part of that resilience is a studious avoidance of knowledge of or dependence on higher layers. This is possible only by means of rigorous standards that define a stable and virtually opaque behavior for each layer relative to the next layer above. For example, IP packets can be carried inside either ethernet or WiFi packets because each has a well-defined standard interface for an IP router to follow. The IP routing does not depend on whether the medium is copper or fiber and in turn presents the same interface for HTTP or FTP control protocols. Each higher layer on the interoperability stack moves farther away from hard criteria of successful exchange between physical components and farther towards soft or

Human-centric

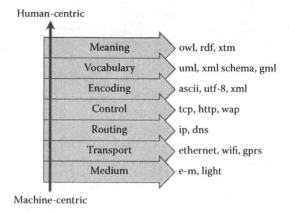

Meaning	owl, rdf, xtm
Vocabulary	uml, xml schema, gml
Encoding	ascii, utf-8, xml
Control	tcp, http, wap
Routing	ip, dns
Transport	ethernet, wifi, gprs
Medium	e-m, light

Machine-centric

FIGURE 8.1   Basic interoperability stack for distributed computing.

abstract criteria of successful communication between people. At each layer, though, simple and stable standards have enabled the next layer to be constructed in a way that decouples it from concerns of lower layers and facilitates a focus on a particular elaboration of interoperability.

At the highest level shown in figure 8.1, the focus is on higher aspects of interoperability, such as semantic interoperability or resilience of meaning to heterogeneous vocabularies. It seems reasonably straightforward to relate a term in one vocabulary to a term in another and thereby enable either to be exchanged without disrupting communication of meaning. Of course few vocabularies are so simple in detail, but the implication here is that of a shared meaning or conceptualization that can be communicated (Gruber 1993). In all knowledge, but particularly in the case of spatial knowledge, there is really very much more detail to be fleshed out in forming the relationship between one person's conceptualization of the physical world and another's. Often such worldviews are indeed shared and semantic interoperability is enough, but subtle heterogeneities can also disrupt effective communication as surely as a broken cable.

Some of the additional levels of concern particularly relevant to spatial information exchange are shown conceptually in figure 8.2, which depicts above all the significant distance between simply exchanging geographic terms and truly communicating fundamental modes of human engagement with the world. At the level of ontology, common vocabularies are the medium of exchange. At the level of representation, the commonality concern is the geometric representation of geographic features. At the application level, a feature representation such as an area centroid point may be appropriate for one application (map), but not for another (navigation to a location within the area). Even a common representation is not useful when there is no shared discernment of the feature to be represented: one person's open space is another person's collection of micro-habitats. Commonly discerned features may nonetheless arise from differing theories of reality: is the essential state of a river the state of the water within it or the rate and path by which water flows along it? Heterogeneity may arise more fundamentally from differences in perception: view sheds are not likely to be very meaningful to someone who is blind.

Even the top level shown, that of "Intention" or personal motivation for form-ing and communicating spatial understanding, may seem irrelevant to geosemantics but can be essential in establishing a context for useful geosemantic relationships between heterogeneous terms (Brodaric 2007). It helps explain why the question perhaps most frequently frustrating the dissemination of geodata is something simi-lar to, "who are you and what do you intend to do with my data?" In other words, how disparate are the intentions for data consumption from those that informed data creation? If you intend to land a plane using data from a handheld GPS, for example, you'd better have your insurance paid up and expect little sympathy from whomever digitized the runway outline as a background map detail.

There is a rough correlation of some of these layers, particularly Intention, with the Semantic Web's layer of Trust (Koivunen and Miller 2001), as well as a long history of logical and philosophic inquiry in this area, which is fascinating but unfortunately beyond the scope of this chapter (see Sowa 2000). Here, the practical consequences for communication of spatial information are our main concern.

8.3.2 Pain Points

On the one hand, each of the layers illustrated in figure 8.2 carries the potential for heterogeneity and disruption of communication. On the other hand, each presents an opportunity for geosemantic standards to introduce a degree of uniformity or at least disclosure, which can mitigate this potential. For example, Village and Town features may use differing terms, but a standard ontology of administrative units can allow them to be exchanged as subtypes of municipality. One town feature may have a point geometry and another town feature a polygon geometry, but standards such as the General Feature Model (ISO 2003a) allow them to be compared as discerned features despite differing representations. Numerous examples may be drawn for the other levels presented in figure 8.2.

This leads us back to consideration of where and how to draw the distinction between SII and spatial applications built upon it. Setting aside for a moment the fact that an "Application" layer is shown explicitly in the figure 8.2 stack, we might fruitfully

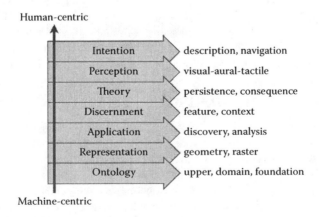

FIGURE 8.2 Expansion of the interoperability stack for exchange of meaning.

ask what interoperability/heterogeneity pains are commonly and repeatedly encountered in geocomputing and which seem more dependent on the particular application to which geocomputing is being applied. For example, discovery of geodata is commonly plagued by geosemantic heterogeneity in both feature and attribute vocabularies (Lieberman 2006). Query and mediation of geodata are additionally challenged by representational heterogeneity, such as between city boundary polygons and city center points. Reuse and composition of geoservices (Lemmens 2006) is almost always complicated by the lack of "is-a" relationships usefully defined between discerned features, for example, "lane" is still a type of street that can be used in a geocoding or routing workflow. On the other hand, problems of cognitive heterogeneity, or two theories for the same real-world entity (is it a steep meadow, or a ski slope?), are largely beyond the concerns of most semantic tools and are painful mainly in specific applications where both models happen to come into play. As an initial hypothesis, then, vocabulary matching, geometric representation of geographic entities and feature discernment are candidates for inclusion into SII, whereas issues of cognition and intent appear for now to be more tractable on smaller application and community scales.

8.4   THE CASE FOR GEOSEMANTIC STANDARDS

If it is the case that we have both higher expectations of what information infrastructure should do for us, and a larger/higher sense of the common functionality that it would be reasonable to consider as infrastructure, then the foregoing shows that geographic information standards will be required to support them. Standards are needed to support greater discrimination in the transmission and retrieval of information. Other standards are required to enhance local discovery and interpretation of remote information. Geosemantic standards, that is, standards for formal semantic representation and logical processing of spatial information, are indicated for both of these roles. Certainly other sorts of standards will be necessary; it is equally true that not all geosemantic standards or practices will be appropriate. There is an important distinction to be made between those simple, basic standards or practices that are appropriate for SII and those that are more appropriate to specific applications supported by SII. In terms of the layers of interoperability discussed above, geosemantic standards can be helpful, on the Web or elsewhere, not only in terms of vocabulary matching, but also to support higher layers of information interoperability, each of which have their unique spatial aspects (Bishr 1998). Other research and technology development may be able to address these levels more flexibly and generally in the future; for now, standards are essential.

8.5   GEOSEMANTIC STANDARDS METHODOLOGY

Various activities within the Open Geospatial Consortium (OGC), including the newly renamed Geosemantics Working Group, as well as the Geospatial Incubator activity within W3C (GeoXG Wiki 2006), have been looking into just this question of what basic geosemantic standards are needed to develop spatially "intelligent" SII. One continuing issue of what to standardize concerns the ambiguous distinction between data and metadata. For example, it may be simpler to start with standards for metadata geosemantics and geosemantic processing without attempting to convert existing data

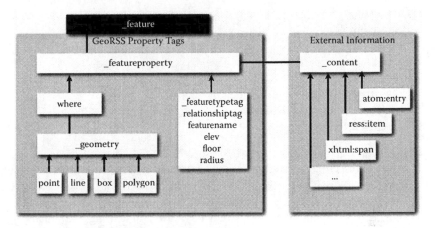

FIGURE 8.3 The GeoRSS content model.

themselves. There are clearly limits to this approach, however (Lieberman et al. 2006). Metadata may be "data about data," but in another sense, it may also be termed "statements about statements." For one application, the former statement may only concern management (e.g. timestamp of last update or service endpoints of repository locations). For another application, that statement may add critical properties to existing statements that completely alter its interpretation without at all changing the target statements themselves.

A critical developing standard, GeoRSS (2006) performs this function by defining a geo-tag vocabulary that can add simple yet formal spatial and geosemantic representations to any of the vast quantities of implicitly spatial information already existing online. The abstract content model for GeoRSS is shown in figure 8.3. For many applications, the GeoRSS tags are simply another form of descriptive metadata, much like a publication date or author name. Looked at as a geosemantic vocabulary, however, GeoRSS provides a way of completely redefining the description of a resource within the ISO/OGC General Feature Model (ISO 2003a) by reference, without modifying the original resource itself. The many uses of this lightweight approach for geosemantically enabling existing information are just now being explored. Past the initial step of geo-enabling information resources with metadata, however, applications are being developed to perform geosemantic processing on the tagged resources as "new" information. This presents a motivation for development of additional geosemantic standards.

8.6   OTHER GEOSEMANTIC STANDARDS

A number of useful categories of basic geosemantic standards for future development have been identified by the W3C (GeoXG Wiki 2007). These categories have subsequently been expanded somewhat and used to characterize existing spatial ontologies (NGA 2007). They are generally characterized here for geosemantic purposes as ontologies to be serialized at least in OWL, but it is clear that they also represent fundamental parts of the most common models for working with geographic information:

1. Geo ontology
2. Feature ontology
3. Feature type ontology
4. Spatial relationship ontology
5. Toponym ontology
6. Coordinate reference/spatial index ontology
7. Geodata set/metadata ontology
8. Spatial services ontology

These basic standards are discussed in greater detail below.

Geo ontology: The geo ontology was first defined in 2003 as latitude and longitude properties of a point (Brickly 2003). The current update of this ontology (Geo 2007) adds a realization of the GeoRSS content model (updated from GeoRSS 2006) that conforms to the General Feature Model as well as including additional geosemantically expressive properties such as featuretypetag, while remaining small and lightweight. An example of the geo:line element from this ontology used within an RSS 1.0 feed item is shown below.

```
<?xml version='1.0' encoding='UTF-8'?>
<rdf:RDF xmlns:rdf='http://www.w3.org/1999/02/22-rdf-syntax-ns#'
    xmlns:geo='http://www.w3.org/2005/Incubator/geo/XGR-geo-owl-2007#'
    xmlns='http://purl.org/rss/1.0/'>
  ...
    <item rdf:about='http://example.com/geo'>
      <title>A walk in the park</title>
      <link>http://example.com/geo</link>
      <description>Just an example</description>
      <geo:line>40.73158   -73.999559   40.732188   -73.999079
40.732688 -74.0234</geo:line>
      <geo:featuretypetag>footpath _ centerline
</geo:featuretypetag>
    </item>
  ...
</rdf:RDF>
```

Feature ontology: A more complete geosemantic realization of the General Feature Model for the representation of information directly as features has not yet been completed, although a number of useful initial steps have been taken, either from the ISO model or from Geographic Markup Language schemas (Islam et al. 2004).

Feature type ontology: While the feature ontology would provide the basic building blocks for defining features, the feature type ontology would then use those blocks to standardize agreement at many levels on basic features such as road, building, lake, etc., the 50 to 100 types that almost everyone could agree upon. One example of a feature type ontology for administrative boundaries has been developed by the U.K. Ordnance Survey (2007).

Spatial relationship ontology: There are a number of commonly accepted and used two-dimensional topological relationship schemes, such as RCC8 or Egenhofer relations, some of which have been developed into partial spatial relationship ontologies, but an additional goal of this ontology would be to provide a means of representing (in more actionable terms) some more common informal relationship expressions such as "across the way."

Toponym ontology: Place names are an important means of geo-locating resources, at least to some approximation. Some work has been done on globally useful place name ontologies (Geonames 2006), but it has not yet really been brought together as a common reference and dual representation of location (together with geometric coordinates).

Coordinate reference and spatial index ontology: One of the fascinating aspects of putting geography on a map is that there are so many different ways of doing it and all of them are in one way or another "wrong." Heterogeneous usage of coordinate reference systems (CRS) and grid/spatial index systems is probably the most frequently occurring spatial heterogeneity problem. It is also the flip side (TCP-IP to DNS) of place names. Although WGS84 might be sufficient globally, other CRSs are important for local geography or different views of the globe (e.g. polar). A useful quantitative way of both geo-locating and indexing resources involves identifying the relevant cell of a spatial index scheme (pyramid of successively smaller cell sizes). A correspondence between grid scheme and map tile scheme would also allow map tiles to easily be discovered along with co-located resources.

Geodata set and data product ontology: This has also been termed the geo-metadata ontology, but given the discussion above on both the integral nature of metadata and the importance of supporting higher levels of spatial interoperability with clear standards, this can also be characterized as the vocabulary for distinguishing collections or products of feature instances (e.g. by creator). Such distinctions can then be acted upon in higher layers of the stack or in specific applications. An example of such an ontology derived from ISO metadata standard 19115:2003 (ISO 2003b) has been developed by Islam et al. (2004).

Spatial Services ontology: Evolving standards such as OWL-S have raised the bar on formal and actionable descriptions of Web services, but elaborations are needed of the manner in which especially the closely coupled content of spatial Web services affects process model and behavior (Lemmens 2006). Lemmens describes in chapter 7 of this book the requirements, characteristics and usage of such an ontology in detail.

8.7   CONCLUSION

Each of these spatial ontology categories already contains at least preliminary examples of ontologies and is a ripe field for research and development of advanced capabilities and applications in many domains. There is some basic structural commonality in each, however, which provides a common foundation for further work. This basic standardization is also critical to achieving an SII that meets our elevated expectations for capability, not only delivering information but truly facilitating its use and reuse without being overcomplicated. Finally, an SII remains ultimately limited in its usefulness if it does not support the evolution of a geosemantic counterpart to the emerging Semantic Web; yet geosemantic interoperability particularly depends on standards—at least for the present—which are able to account for many of the levels at which people imbue their physical world with meaning. Geosemantic standards are not completely "hopeless without," but it is reasonable to argue that they are at least quite urgently "nice to have."

REFERENCES

Bishr, Y. 1998. Overcoming the Semantic and Other Barriers to GIS Interoperability. *Int. J. Geographical Information Science* 12 (4):299–314.

Brickley, D., ed. 2003. Basic Geo Vocabulary. http://www.w3.org/2003/01/geo/.

Brodaric, B. 2007. Geo-Pragmatics for the Geospatial Semantic Web. *Transactions in GIS* 11 (3):453–477.

Geo. 2007. Worldwide Web Consortium Geospatial Incubator. Final report. Available at http://www.w3.org/2005/Incubator/XGR/.

Geonames. 2006. Geonames Ontology. http://www.geonames.org/ontology/.

GeoRSS. 2006. GeoRSS Content Model. http://www.georss.org/model.

GeoXG. 2006. Worldwide Web Consortium Geospatial Incubator. http://www.w3.org/2005/Incubator/geo.

GeoXG Wiki. 2007. Worldwide Web Consortium Geospatial Incubator Wiki. http://www.w3.org/2005/Incubator/geo/Wiki/Spatial_Ontologies.

Gruber, T. 1993. A Translation Approach to Portable Ontology Specifications. *Knowledge Acquisition* 5 (2):199–220. Also available online at http://ksl-web.stanford.edu/knowledge-sharing/papers/README.html#ontolingua-intro.

Hart, G. 2006. So What's So Special about Spatial? Paper presented at Terra Cognita workshop, International Semantic Web Conference, Atlanta, GA, November 2006.

Islam, A. S. et al. 2005. List of OWL Ontologies based on Norms. http://loki.cae.drexel.edu/~wbs/ontology/list.htm and links therein.

ISO. 2003a. ISO IS 19109:2003, Geographic Information—Rules for Application Schema. Geneva: International Organization for Standardization.

ISO. 2003b. ISO IS 19115:2003, Geographic Information—Metadata. Geneva: International Organization for Standardization, August 2007.

Koivunen, M., and E. Miller. 2001. W3C Semantic Web Activity. http://www.w3.org/2001/12/semweb-fin/w3csw.

Lemmens, R. L. G. 2006. Semantic Interoperability of Distributed Geo-Services. Ph.D. thesis, International Institute for Geo-Information Science and Earth Observation (ITC), Enschede, The Netherlands.

Lieberman, J. et al. 2006. *Geospatial Semantic Web Interoperability Experiment Report*. Discussion Paper 06-002r1. Wayland, MA: Open Geospatial Consortium Inc.

NGA. 2007. *Geospatial Ontology Trade Study*, ed. J. Ressler and M. Dean. Contract HM1582. Bethesda, MD: National Geospatial-intelligence Agency. Also available from http://projects.semwebcentral.org/docman/?group_id=84.

Ordnance Survey. 2007. Administrative Geography Ontology, http://www.ordnancesurvey.co.uk/oswebsite/ontology/.

Sowa, J. 2000. Ontology, Metadata, and Semiotics. In *Conceptual Structures: Logical, Linguistic, and Computational Issues*, ed. B. Ganter and G. W. Mineau, 55–81. Lecture Notes in AI no. 1867. Berlin: Springer-Verlag.

9 A Standardized Land Administration Domain Model as Part of the (Spatial) Information Infrastructure

Arco Groothedde, Christiaan Lemmen,
Paul van der Molen, and Peter van Oosterom

CONTENTS

Spatial data sets are most useful in the support of decision making, management of space, performance of government and business, etc., when integrated in governmental information infrastructures (architectures). This implies availability of well-maintained links between spatial data sets and other basic or key data sets, for example, on addresses, persons, companies, buildings, land rights, etc. Integrated and interorganizational value chains, business process management, and reduction in administrative overheads can be introduced based on new business models. In general, the resolution of problems in society requires more information than provided from one single data set, and this is equally true for problems with a spatial concept. It is evident that this type of data provision is complex in cases where data are stored at a variety of locations and in data models specific to their application domains. In this chapter it is argued that an effective infrastructure can be achieved solely by the use of authentic registers (or "key registers") to store key data that are available for integration and multiple use.

A standardized Land Administration Domain Model (LADM) provides an extensible basis for efficient and effective cadastral system development based on a model-driven architecture (MDA), and enables involved parties to communicate based on the shared ontology implied by the model. As it is already difficult within one domain (such as land administration) to agree on the concepts used and their semantics, it will be even more difficult in dealing with other domains. However, we cannot avoid this if a meaningful interoperable spatial information infrastructure has to be developed and implemented.

In this chapter we discuss standardization at the domain level with land administration as a case (section 9.1). Programs at European and national government levels are signaling opportunities, setting policies and taking measures to capture the benefits of information and communication technology (ICT). In tandem with societal demands, this will lead to government services via the Internet, key registers, Web services and more. Developments on (S)II at the national (with The Netherlands as a case) and European scales are introduced in key registers in The Netherlands (section 9.2). The effects of the (S)II on registration will then be analyzed in effects of (S)II on a registration (section 9.3). An introduction to the standardization efforts on the cadastral domain is provided in standardization of the cadastral domain (section 9.4). Impact of further integration of cadastral data onto a spatial information infrastructure is discussed in section 9.5.

9.1   STANDARDIZATION AT THE DOMAIN LEVEL

Sensing technologies, Global Navigation Satellite Systems (GNSS) and wireless communication have improved gathering and use of information on the Web, resulting in a worldwide increase in the use of digital geographic data. This has led in turn to renewed interest in applications employing geographic data, how objects relate spatially and new geo-information dissemination possibilities such as Google Earth, maps.com and Microsoft Virtual Earth. There are fast-growing possibilities for online use of geographic data for all kinds of analysis, and the reliance of society upon such data is growing commensurately. To enable the use of data from multiple national and international data sources, a worldwide structure must be developed for describing digital geographic data and services. This is the aim of the International Organization for Standardization (ISO), in close cooperation with other organizations such as the Open Geospatial Consortium (OGC). Using ISO standards, a national standard for the exchange of geodata sets based on a semantic model is currently being implemented in The Netherlands using the base model geo-information of The Netherlands. For cross-border access to geodata, a European metadata profile based on ISO standards is under development using rules of implementation defined by the Infrastructure for Spatial Information in the European Community, INSPIRE. For actual data exchange, the INSPIRE implementing rules will further define harmonized data specifications and network services. This is complemented with data access policies and monitoring and reporting on the use of INSPIRE (see chapter 1 describing INSPIRE in more detail). From the Kadaster point of view, the main areas for Web services development include business-process management, data acquisition, online use of data from multiple sources, and e-commerce applications (Groothedde 2006).

In this chapter the role of a cadastral registration (or land administration) as an important component of the (spatial) information infrastructure—(S)II—is investigated. In our view, the cadastral registration itself contains both spatial information, for example, on land parcels, and administrative information, for example, registered real rights. In addition, the cadastral registration has important relationships with other registrations in the (spatial) information infrastructure, again both spatial, for example, topography, buildings and administrative information: persons, addresses, companies/trade. It is therefore important to have unambiguous definitions of the contents of these registrations in order to avoid overlap and to enable reuse of information in other registrations. Further, due to continuous updating of these independent, but related, registrations, care has to be taken to maintain consistency, not only within one registration, but also between registrations in the information infrastructure. By reusing basic standards (geometry, temporal, metadata, observations, and measurement), at least the semantics of these fundamental parts of the model are shared and well defined. What is needed in addition is domain-specific standardization to capture the semantics of the cadastral domain on top of this agreed foundation (ISO General Feature Model). The model should be specified in a Unified Modeling Language (UML) class diagram and then converted into an eXtended Markup Language (XML) schema, which can then be used for actual data exchange in our networked society (interoperability).

9.2   (S)II DEVELOPMENTS: KEY REGISTERS IN THE NETHERLANDS

Information and communication technology (ICT) offers many opportunities for improving the performance of government and business. Areas that may profit include education, safety, health care, international cooperation, economic efficiency (integrated value chains, business-process management, and reduction in administrative overheads), prevention and detection of fraud, and accident and disaster management. ICT trends such as ubiquitous access, smart objects, open source, increased bandwidth, interoperability, and data-exchange standards will result in new business models. New perspectives are opened up by options like increased location independence, high-quality online services based on immediate access to all required data, use of identified objects available for process control, integration within business chains and government organizations, and increased e-shopping.

The basic idea behind information infrastructures is that they provide for tools giving easy access to distributed databases to people who need those data for their own decision-making processes. Although information infrastructures have a substantial component of information technology, the most fundamental asset is the data themselves (commonly agreed on representations of real-world phenomena or social constructs), because without data there is nothing to have access to, to be shared or to be integrated. In the last decade it was understood that the development of information infrastructures not only provided easy access to distributed databases, but also gave good opportunities for rethinking the role of information supply for the performance of governments. Based on this starting point, the "Streamlining Key Data" Programme of The Netherlands government (van Duivenbode and de Vries 2003) took the lead in the development and implementation of a strategy for restructuring government information. This is done in such a way that an electronic government evolves that

- Inconveniences the public and the business community with requests for data only when this is absolutely necessary
- Offers them a rapid and good service
- Cannot be misled
- Instills the public and the industrial community with confidence
- Is provided at a cost that is not higher than strictly necessary

Jointly with a number of other government registers (see figure 9.1), the property registers and cadastral maps and topographic maps of The Netherlands Cadastre, Land Registry and Mapping Agency (hereinafter called Kadaster) have been formally appointed in 2002 as "key registers" of the governmental information infrastructure. The key registers will be the core of a system of so-called authentic registers, which might be any register that is maintained by a single government body and used by many others as the authentic source of certain data. If a register is formally designated as an authentic register, all other government organizations are strictly forbidden to collect the same data by themselves. In their budget allocation they will not find any money for data collection at this point. The "Streamlining Key Data" Programme concentrates on two goals:

- The communal use of data—in principle data would be collected on one occasion and used repeatedly for the implementation of series of laws and regulations.
- The joint use of data—data from different records required for the performance of a specific government duty would be combined in one database.

An authentic register is defined in the program as "a high quality database accompanied by explicit guarantees ensuring for its quality assurance that, in view of the entirety of statutory duties, contains essential and/or frequently-used data pertaining to persons, institutions, issues, activities or occurrences and which is designated by law as the sole officially recognised register of the relevant data to be used by all government agencies and, if possible, by private organizations throughout the entire country, unless important reasons such as the protection of privacy explicitly preclude the use of the register" (van Duivenbode and de Vries 2003).

The resolution of problems in society often requires more information than provided from one single data set, and this is equally true for problems with a spatial concept. It is evident that this type of data provision is complex when data are stored at a variety of locations and in data models specific to its application. An effective infrastructure can be achieved solely by the use of authentic registers (or "key registers") to store key data that are available for integration and multiple use. Various countries work on this subject. The Streamlining Key Data Programme offers for The Netherlands the appropriate policy. This is to the benefit of efficient and effective performing authorities, and contributes to the reduction of the administrative overhead in both the public and business environments. Legislation is currently

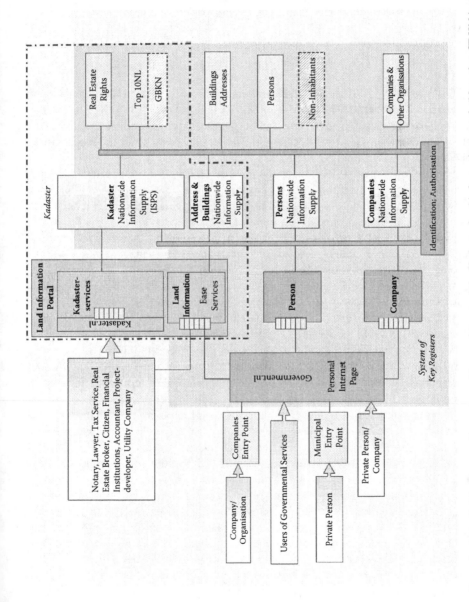

FIGURE 9.1 (See color insert.) A landscape: the system of key registers and the Kadaster land information portal. (Note that GBKN is the Dutch acronym for *Grootschalige BasisKaart Nederland* or, in English, Netherlands Large Scale Topographic Base Map 1:1,000.)

being prepared—or has been approved—for the confirmation of the designation of the following registers:

- Municipal Personal Records Database–Population Register
- Cadastre (Parcels and Rights)
- Company Key Register ("New Trade Register")
- Addresses
- Buildings
- Topography (TOP10NL)

On February 8, 2007 the Dutch Parliament approved the Act on Basic Registration Cadastre and Topography. The implementation date was January 1, 2008. The Municipal Personal Records Database has also been accepted as an authentic register; the acts where Buildings and Addresses and, further, the New Trade Register will be appointed as key registers are in process.

Experience acquired with the Municipal Personal Records Database (the population register, which cannot yet be consulted online) indicates that the Kadaster could play a role in rendering these addresses and buildings accessible at a national level, even though the municipalities remain the owner of the source. The Kadaster's justification for this approach is based on one of the agency's core competences, that is, its skills in the management and maintenance of national databases with an extremely high update frequency.

It is Kadaster's strategy to play a leading role in the system of key registers. Figure 9.1 provides an overview of the system of key register.

One can observe that this infrastructure does not only concern spatial data. Kadaster will review the extent to which supplementary relevant data could be included in the land register. The Kadaster can play a leading directive role in the organization of the provision of this information to the market players, whereby consideration will need to be given to cooperation with some registers within the context of digital availability and fast accessibility. The Kadaster can acquire a good position by the provision of a series of topographic and geographic products that possess an internal consistency and are indispensable to third parties within the context of spatial planning, land use, management, and maintenance. For this reason the cadastral map, the Large Scale Topographic Base Map 1:1,000 and the Topographic Key Register 1:10,000 will need to be object-oriented and maintained mutually consistent by means of data set integration using ontologies. Advanced detection of changes, for example, using satellite images followed by the processing of the changes in all data sets ("change propagation"), will then become a feasible proposition. The assumption of the management of, for example, the General Elevation Dataset of The Netherlands, and the National Road Database, indispensable to dynamic traffic management, would be compatible with this. The integration of the National Triangulation (RD*: x and y) and Elevation Datum (NAP†: z) in a 3D reference frame would result in a pivotal role in the geometric infrastructure, inclusive of elevations.

---

* RD is the Dutch acronym for *Rijksdriehoeksmeting.*
† NAP is the Dutch acronym for *Normaal Amsterdams Peil.*

At the European level, the priorities include completion of a Single European Information Space promoting an open and competitive internal market for information society and media, strengthening research into innovation, and investment in ICT, and achieving an inclusive European information society. Objectives have been set for improving the security and reliability of broadband services, for creating better and smarter use of ICT within the public domain and for improving interoperability. Another issue at the European level is better access to (key) data sets. The Infrastructure for Spatial Information within the European Community (INSPIRE) aims to create a system for access to and exchange of spatial information for environmental monitoring. Cadastral and topographic data are considered relevant environmental data and will thus be included in discussions on European harmonized (meta)data content, network services, data access policy, and monitoring data/services use.

Based on the above, Kadaster's current strategic objectives might be reformulated as aiming for the best possible performance of current public duties and promotion of innovation and knowledge for the adoption of a leading role in their evolution in response to societal developments. Strategic subobjectives include

- Investigation of evolution towards a (more) positive land registration system
- Introduction of a 3D land register
- Ambition to adopt a role as a center for a range of key registers
- Provision of more complete insight into private and public legal status of registered property
- Achieving a substantial role in organizing information needs of the property market chain
- Provision of an appropriately linked set of topographic and geographic data sets, object-oriented and mutually consistent with respect to change
- Fulfillment of a pivotal role in geometric infrastructure (x, y and z)
- Acceptance of a prominent EU partner role in harmonizing registered property law, land registration, and cadastres
- Development of flexible land-planning instruments suitable for use in realizing a variety of societal spatial objectives

9.3 EFFECTS OF (S)II ON A REGISTRATION

In this section attention is paid to three aspects that require specific attention when a certain registration is playing a role in the (S)II. The first observation is that the information content within the (S)II consists of several registrations and that it is therefore important to define what content belongs to which registration, that is, defining the boundaries of the registrations. In the first part of this section this will be discussed in detail for the cadastral registration (also called land administration). The second observation is that the different registrations are related, that is, there are references in the content from one registration to another registration. As the registrations are maintained by largely autonomous organizations, care has to be taken when information is updated that related registrations are informed (in order to trigger potential related updates elsewhere). This topic will be discussed in the last part of this section. The third and final observation is that an unambiguous data specification is of course needed for the registration itself, but also when addressing the first two issues mentioned above (registration boundary and consistency between registrations after updates). How to achieve harmonized data specification of a registration (both with respect to other domain registrations in the same country and with respect to the same domain in other countries) is also introduced in this section. The next section focuses on the standardization of the cadastral domain.

First is the issue of the content of a specific registration, and for this the cadastral registration will be used as an example. The result of comparing current cadastral registrations in different countries depends a lot on the equal scope of the models; for example, if one cadastral model includes a person registration (with all attributes and related classes to persons) and the other model just refers to a person (in another registration), then the two models may look different, but the intentions are the same. Only the system boundary of the involved models is different. It is therefore proposed to try to get some consensus on the model boundary by considering the current cadastral registration practice in different countries of the world. Next is an attempt to list themes that are related to, but outside, the Land Administration Domain Model (see the next section for more details):

1. Spatial (coordinate) reference system
2. Ortho photos, satellite imagery and Lidar (height model)
3. Topography (planimetry)
4. Geology, geotechnical, and soil information
5. (Dangerous) pipelines and cable registration
6. Address registration (including postal codes)
7. Building registration, both (three-dimensional) geometry and attributes (permits)
8. Natural person registration
9. Non-natural person (company, institution) registration
10. Polluted area registration
11. Mining rights registration
12. Cultural history, (religious) monuments registration
13. Ship and airplane (and car) registration

The first four topics listed above are or can be used in the cadastral system for reference purposes (or for support in data entry). Other topics have a strong relationship in the sense that these (physical) objects may result in legal objects ("counterparts") in the cadastral registration. For example, the presence of utility cables or pipelines can also result in a restriction area (two- or three-dimensional) in the cadastral registration. However, it is not the cable or pipeline itself that is represented in the cadastral system; it is the legal aspect of this. Though strongly related, these are different aspects. Compare this to a wall, fence, or hedge in the field and the "virtual" parcel boundary. The fact that these "external" objects (or packages) are so closely related also implies that it is likely that some form of interoperability is needed. When the cables or pipelines are updated, then both the physical and legal representations should be updated consistently (within a given amount of time). This requires some semantic agreement between the "shared" concepts or at least the interfaces and object identifiers. In other words, these different but related domain models need to be harmonized. As it is already difficult within one domain (such as the cadastral world) to agree on the concepts used and their semantics, it will be even more difficult when we are dealing with other domains. However, we cannot avoid this if a meaningful, interoperable geo-information infrastructure is to be developed and implemented. It seems appropriate also that a more neutral organization plays a coordinating role in this harmonization process, such as the OGC, ISO, INSPIRE, FIG (International Federation of Surveyors), or CEN (European Committee for Standardization).

In several countries of the world we see attempts to harmonize a number of domain models within one country, for example, Australia, Germany, The Netherlands. But this is not sufficient, as the models should also be harmonized internationally as in the case of INSPIRE. One could raise the question: what is the best order for harmonizing, first within a specific domain (at an international level as, for example, is the case with the LADM) and then harmonizing these different domains, or first within a specific country (including all relevant domains) and then harmonizing these different country models? Anyhow, it will be an iterative process as our insight and knowledge will keep on being refined (and both approaches will probably be applied). An extremely important aspect of the future (S)II, in which (related) objects can be obtained from another source (instead of copied), is that of "information assurance." Though the related objects, for example, persons in the case of a cadastral system, are not the primary purpose of the registration, the whole cadastral "production process" (both update and delivery of cadastral information) does depend on the availability and quality of the data at the remote server. Some kind of "information assurance" is needed to make sure that the primary process of the cadastral organization is not harmed by disturbances elsewhere. In addition, remote (or distributed) systems/users might not only be interested in the current state of the objects, but they may need an historic version of these objects, for example, for legal claims, taxation, or valuation purposes. So even if the organization responsible for the maintenance of the objects is not interested in history, the distributed use may require this (as a kind of "temporal availability assurance").

Finally, a fundamental question is: how to maintain consistency between two related distributed systems in case of updates? Assume that system A refers to object X in system B (via object id B.X_id). Now the data in system B are updated and object X_id is removed. As long as system A is not updated the reference to object X should probably be interpreted as the last version of this object available. Note that the temporal aspect is again getting a role in and between the systems! The true solution is of course also updating system A and removing the reference to object X (at least at the "current" time). How this should be operationalized will depend mainly on the actual situation and involved systems. It might help to send "warning/update messages" between systems, based on a subscription model of the distributed users/systems.

9.4 STANDARDIZATION OF THE CADASTRAL DOMAIN

In order to obtain an unambiguous definition of the content of a cadastral registration, at the FIG Congress in Washington, D.C. in 2002, the proposal was launched to develop a (shared) Core Cadastral Domain Model: the FIG CCDM (Lemmen and van Oosterom 2006), which has recently been renamed the Land Administration Domain Model: (LADM), as in some contexts the term CCDM caused confusion and misinterpretations. After the launch several specific international workshops have been devoted to the development of this topic and various organizations have been involved (the OGC, International Organization for Standardization, ISO/TC211, UN-Habitat, INSPIRE); M.Sc. and Ph.D. students, researchers and international experts have devoted a significant part of their research to cadastral modeling, resulting in a series of versions of the CCDM/LADM published in different magazines, proceedings and journals.

A standardized LADM, covering land registration and cadastre in a broad sense, the "multipurpose cadastre" (Kaufmann and Steudler 1998), serves at least two important goals: (1) to avoid reinventing and reimplementing the same functionality over and over again, but provide an extensible basis for efficient and effective cadastral system development based on a model-driven architecture (MDA); and (2) enable involved parties, both within one country and between different countries, to communicate based on the shared ontology implied by the model. The second goal is very important for creating standardized information services in an international context, where land administration domain semantics have to be shared between countries (in order to enable needed translations). But the second goal is also important within one country, in order to meaningfully combine and exchange information from several different registrations in the information infrastructure.

Important conditions during the design of the model were and still are: it should cover the common aspects of cadastral registrations all over the world, it should be based on the conceptual framework of Cadastre 2014 (Kaufmann and Steudler 1998), it should follow the international ISO and OGC standards, and at the same time be as simple as possible in order to be useful in practice. The LADM itself represents an important new wave in geo-information standardization: after the domain-independent basic geo-information standards (current series of ISO and OGC standards), the new standards based on specific domains will now be developed. Due to historical differences between countries (and regions) similar domains, such as the land administration domain, may be modelled differently and therefore nontrivial harmonization has to be done first. The LADM is a result of this harmonization and one of the first presented examples of semantic geo-information domain standards.

A cadastral parcel (figures 9.2, 9.3, and 9.4) is single area of land or more particularly a volume of space, under homogeneous real property rights and unique ownership (UNECE 2004; WG-CPI 2006). By unique ownership is meant that the ownership is held by one or several owners for the whole parcel. By homogeneous property rights is meant that rights of ownership, leases, and mortgages affect the whole parcel. This does not apply to specific rights as servitudes, which may only affect part of the parcel. Irrespective of the legal system adopted by each member state, the cadastre is defined as a register under the responsibility of the government. Its use complies with the principles of equality, security, and justice to all the citizens of the European Union. Access to cadastral information is ruled by laws and regulations in order to protect personal information. The classical cadastre basic unit is the parcel. Parcels can be grouped in immovable register objects (figure 9.3).

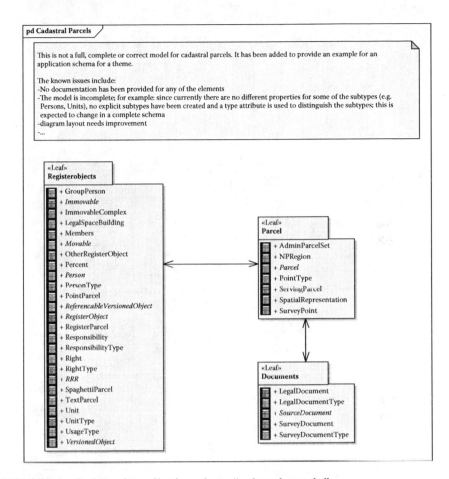

pd Cadastral Parcels

This is not a full, complete or correct model for cadastral parcels. It has been added to provide an example for an application schema for a theme.

The known issues include:
-No documentation has been provided for any of the elements
-The model is incomplete; for example: since currently there are no different properties for some of the subtypes (e.g. Persons, Units), no explicit subtypes have been created and a type attribute is used to distinguish the subtypes; this is expected to change in a complete schema
-diagram layout needs improvement
-...

«Leaf»
Registerobjects
+ GroupPerson
+ *Immovable*
+ ImmovableComplex
+ LegalSpaceBuilding
+ Members
+ *Movable*
+ OtherRegisterObject
+ Percent
+ *Person*
+ PersonType
+ PointParcel
+ *ReferencableVersionedObject*
+ *RegisterObject*
+ RegisterParcel
+ Responsibility
+ ResponsibilityType
+ Right
+ RightType
ı *RRR*
+ SpaghettiParcel
+ TextParcel
+ Unit
+ UnitType
+ UsageType
+ *VersionedObject*

«Leaf»
Parcel
+ AdminParcelSet
+ NPRegion
+ *Parcel*
+ PointType
+ ServingParcel
+ SpatialRepresentation
+ SurveyPoint

«Leaf»
Documents
+ LegalDocument
+ LegalDocumentType
ı *SourceDocument*
+ SurveyDocument
+ SurveyDocumentType

FIGURE 9.2 Packages in application schema "cadastral parcels."

A parcel has a nationwide, unique, real property identifier. The spatial description of the parcels and other cadastral objects should be provided with an adequate degree of accuracy. Descriptive data may include the nature, size, value, and legal rights or restrictions associated with each immovable register object under or over the surface (adapted from PCC 2003). Cadastral parcels cover a territory (regional or nationwide) and there are no overlaps or gaps (in reality). An exception to this rule may be government land (or public domain) not registered within the cadastre—though this is not recommended practice.

Besides various types of ownership, cadastral parcels, or, to be more general, immovable register objects, can be associated with other types of real rights (usufruct, superficies, long lease, etc.), responsibilities or restrictions (figure 9.3). The line where a discontinuity in the specific legal situation occurs is the cadastral boundary. Vertices of this boundary can be marked in the field (or not). In many cases field sketches with survey observations are available as a source document (figure 9.6). Observations (classical surveying: directions or bearings, angles, and distances combined with control points or "GPS-based surveying": coordinates) are used to determine coordinates

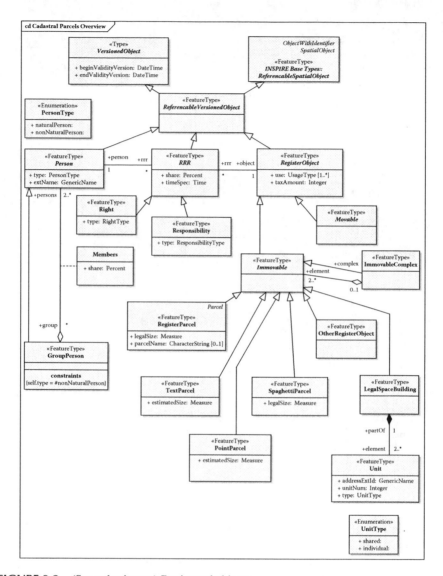

FIGURE 9.3   (See color insert.) Registered objects.

in a projection system; those coordinates are adjusted to the cadastral map. Current practice is to express the coordinates in the cadastral map in the national reference system. In the future this might be changed to the European Terrestrial Reference System (ETRS89), because more and more GNSS (GPS, GLONASS and Galileo) surveys will be used to collect data and this will better enable data consistency near the country boundaries within Europe.

A cadastral boundary does have several attributes of its own. Field sketches (or survey documents) can be used for boundary reconstruction in case of disputes. From a technical point of view, the set of related boundaries is sometimes stored as a closed polygon, with a risk of gaps and overlaps between parcels (this is a quality problem in

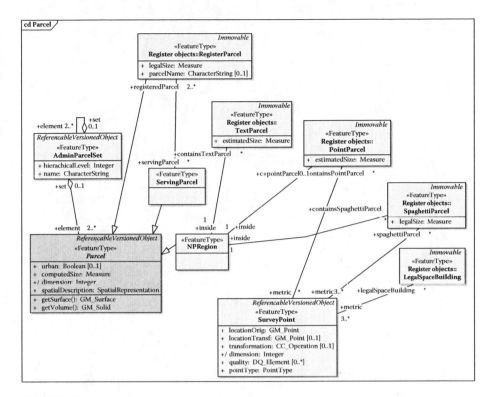

FIGURE 9.4 Parcels.

the database, of course, not in reality). This also implies that every boundary would be stored at least two times (in the "left" and "right" parcels), which is redundant. Further, boundaries also have their own attributes, which have to be attached to a specific instance (which would imply a three representation). In order to avoid these issues, a parcel representation based on a topological structure is often used (figure 9.5). Mostly boundaries do not have a meaningful (based on an administrative hierarchy) identifier, but could be associated with survey documents (which do have some kind of meaningful identifier, known in the outside world).

To illustrate the relationships of the cadastral parcel registration with other registrations within an (S)II, a number of examples from INSPIRE will now be described. Specific boundaries of cadastral parcels are also the boundary of an administrative unit (municipality, province, country); this is an important relationship with theme 4 from Annex I of the INSPIRE directive (Directive 2007/2/EC). Parcels and boundaries have associations with buildings (theme 2 from Annex III of the INSPIRE directive)—sometimes used as a local reference for boundaries, but also used for orientation purposes. Parcels and boundaries have associations with transport networks (theme 7 from Annex I of the INSPIRE directive), for the same orientation purpose, but also roads, railroads, and waterways are separate parcels as they are often owned by government. A strong link exists between cadastral parcels and addresses (theme 5 from Annex I of the INSPIRE directive). Links exist between cadastral parcels, land use (theme 4 from Annex III of the INSPIRE directive) and land cover (theme 2 from Annex II of the INSPIRE directive).

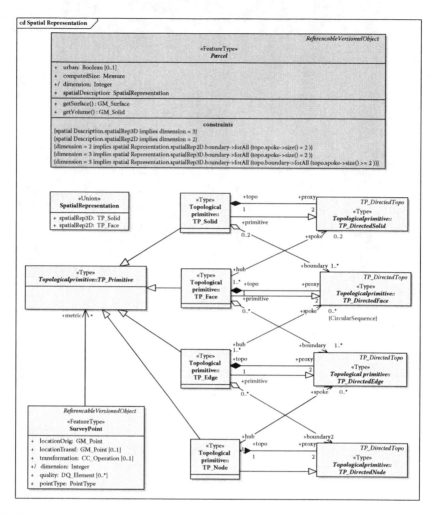

FIGURE 9.5   Spatial representation of parcels and survey points.

Cadastral parcels must have a unique real property identifier to which the legal status is attached. This identifier is often based on a hierarchy of administrative areas (provinces/districts/cantons/..., municipalities/communes/...., sections/polygons/...) and sometimes to the "mother" parcel (subdivision of parcel ..../..../..../37 means for example ..../..../..../37/1 and ..../..../..../37/2). At a European level, the national identifiers should get a country code prefix to make them unique within Europe. Alternatively, there could be explicit associations between predecessors and successors. The cadastral information should be maintained continuously in order to reflect the actual legal situation. Of course, in reality and in information provision there might be a slight delay. Due to the legal importance, the history is currently maintained in some countries, but this may be needed in many countries.

In the appendix to this chapter, the feature catalogue (associated to the UML class diagrams in figures 9.1–9.7) for the cadastral parcel data specification example is given (taken from INSPIRE D2.6 2007) and partly described as an example of how

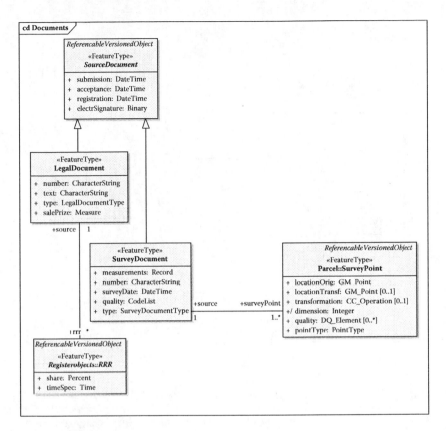

FIGURE 9.6 Documents.

a feature catalogue could look. This feature catalogue is under development and will be based on the ISO 19110 (ISO 2005). As can be observed in the model there are both internal references (e.g. between parcels and boundaries based on the ISO19107 topology model) and external references to information in other registrations of the (S)II, for example, Persons, Buildings, Addresses. The feature catalogue of the cadastral parcel model as presented in the appendix is based on the last version of the LADM (Lemmen and van Oosterom 2006), but adapted to the INSPIRE Generic Conceptual Model (INSPIRE D2.5 2007) and with application of the INSPIRE methodology to derive and describe harmonized data specifications (INSPIRE D2.6 2007). It should be noted that stricter use of the ISO TC211 series of standards has been applied in this version, compared to the previous versions of the model.

9.5 DISCUSSION AND CONCLUSION

Every country (countries can be in a federation) in Europe has a cadastral or land administration system operational (in some countries not yet for the complete territory), often as the responsibility of a national organization, or as the responsibility of a more local government organization. Due to different legal systems and different national traditions, there is a rich variety of cadastral systems around. As this limits

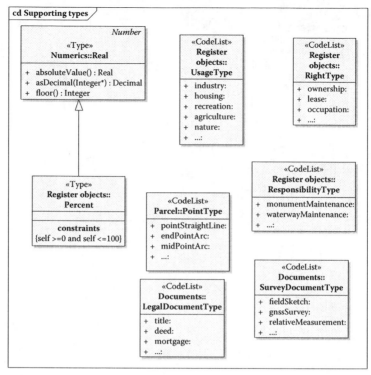

FIGURE 9.7 Supporting types (basic types and code lists).

interoperability (e.g. in the context of EULIS) and results in high system development and maintenance costs, non-governmental (international) organizations, such as the FIG, developed the LADM and submitted this to ISO TC211 as a new work item proposal (N2125). Before the standardization procedure will start, a complete feature catalogue of the LADM will be developed in addition to the UML class diagram.

Cadastres or land information systems form an important part of the land administration systems of the member states. Cadastral activity is related to creating and updating the land parcel's alphanumerical and graphical information and its aggregation. The cadastral organizations in each member state are those public organizations that have specific legal responsibility for creating and updating the land parcel's alphanumerical and graphical georeferenced information or its coordination at the national level (PCC 2003).

Looking at it from a little distance one can observe that the systems are in principle mainly the same: they are all based on the relationships between persons and land, via (property) rights, and are in most countries influenced by developments in ICT. The two main functions of every cadastral system are: (1) keeping the contents of these relationships up to date (based on legal transactions) in a cadastral registration system and (2) providing information on this registration. This chapter has explored which important issues related to a registration in an (S)II have to be confronted (registration content boundaries, keeping related registrations consistent after updates, and harmonized data content). Based on the experience of The Netherlands, Kadaster solutions are proposed to solve these issues and this has been illustrated with on-going standardization activities within several international bodies (such as FIG, ISO TC211, and INSPIRE).

ACKNOWLEDGMENTS

The authors wish to thank the members of the INSPIRE data specification team (especially Clemens Portele for his contribution to the UML class diagram in figures 9.2 to 9.7), and colleagues at Kadaster, with special thanks to Arend Booii. The International Institute for Geo-Information Science and Earth Observation (ITC) and the Delft University of Technology for their support in the preparation of this chapter. Special thanks to Sisi Zlatanova and João Paulo da Fonseca Hespanha de Oliveira for their constructive reviews.

REFERENCES

Directive 2007/2/EC of the European Parliament and of the Council of 14 March 2007, establishing an Infrastructure for Spatial Information in the European Community (INSPIRE). *Official Journal of the European Union*, 25.04.2007, L108/1-14 (Acts adopted under the EC Treaty/Euratom Treaty whose publication is obligatory).

Groothedde, A. 2006. ICT Trends and Institutional Change, with Special Attention to the Land Administration Sector. FIG XXIII Congress, Munich.

INSPIRE D2.6, Drafting Team "Data Specifications." 2007. Methodology for the Development of Data Specifications (draft).

ISO. 2005. ISO 19110 Geographic Information—Methodology for Feature Cataloguing. ISO 19110:2005(E). Geneva: International Organization for Standardization.

Kaufmann, J., and D. Steudler. 1998. *Cadastre 2014, A Vision for a Future Cadastral System.* Copenhagen: International Federation of Surveyors.

Lemmen, C. and P. van Oosterom. 2006. Version 1 of the FIG Core Cadastral Domain Model. FIG XXIII Congress, Munich.

PCC. 2003. Common Principles on Cadastre in the European Union. Declaration, Rome, December 3.

UNECE. 2004. *ECE Guidelines on Real Estate Units and Identifiers.* New York: United Nations.

van Duivenbode, H., and M. de Vries. 2003. *Upstream: Chronicle of the Streamlining Key Data Programme.* The Hague: voormalige programmabureau Stroomlijning Basisgegevens.

WG-CPI. 2006. Role of the Cadastral Parcel in INSPIRE and National SDIs with Impacts on Cadastre and Land Registry Operations. Joint Working Group of EuroGeographics and the PCC (WG-CPI), Inventory document.

APPENDIX

CADASTRAL PARCELS DATA SPECIFICATION

Besides the foundation schemas and the proposed Generic Conceptual Model (see INSPIRE D2.6. 2007) it also contains a draft application schema for a theme, "cadastral parcels." The schema has been included to provide an example for an application schema. The current version of the model previously was shown on the UML class diagrams.

FEATURE CATALOGUE FOR FEATURE-BASED DATA (VECTOR DATA)

The feature catalogue as presented below describes, as a partial example, what a feature catalogue for the Land Administration Domain may look like. A selection of the features as presented in the class diagram (in figures 9.2–9.7) is described here: the RegisterObjects, which can be Movable or Immovable, are introduced. RegisterParcel is one of the specialization classes of Immovable and is presented in this example, as well as ServingParcels and Parcels as specialization classes of RegisterParcel.

FEATURE CATALOGUE METADATA

Feature catalogue name:	cadastral parcels
Scope:	cadastral parcels
Field of application:	multipurpose land administration (ownership, taxation, planning), including social tenures
Version number:	v0.2
Version date:	20-July-2007
Definition source:	INSPIRE DT (EULIS)
Definition type:	full model (feature, attribute, operation, association), example

FEATURE TYPE

Name:	**RegisterObject**
Definition:	Objects that are subject to registration in a (public) registration by law. Contains movable and immovable objects.
Aliases (optional):	Dutch: registerobject
Feature attribute name(s):	use
	taxAmount
Feature association name(s):	to_RRR
Feature operation names(s) (optional):	N.A.
Subtype of:	ReferencableVersionedObject

FEATURE ATTRIBUTE

Name:	use		
Definition:	main use of RegisterObject		
Value data type:	UsageType [1..*]		
Value measurement:	from legal document		
Value domain type:	enumeration type (CodeList)		
Value domain:	Depends on local situation		
Feature attribute value(s):	Label:	Code:	Definition:
	Industry	i	produce goods
	Housing	h	where people live
	Recreation	r	where people play
	Agriculture	a	produce food
	Nature	n	unspoiled environment

FEATURE ATTRIBUTE

Name:	*taxAmount*
Definition:	*amount of real estate tax for the RegisterObject*
Value data type:	*numeric*
Value measurement:	*local currency (UoM, Euro, Pound, etc.)*
Value domain type:	*integer*
Value domain:	*non-negative*

FEATURE ASSOCIATION

Name:	*RegisterObject-RRR*
Inverse relationship:	*RRR-RegisterObject*
Definition:	*Rights, Restrictions and Responsibilities associated with RegisterObject*
Feature types included:	*RegisterObject, RRR*
Order indicator:	
Cardinality:	** at RRR, 1 at RegisterObject*
Constraints:	
Role name:	*rrr, object*

FEATURE TYPE

Name:	**Movable**
Definition:	A movable object
Feature attribute name(s):	
Feature association name(s):	
Feature operation names(s) (optional):	N.A.
Subtype of:	RegisterObject

FEATURE TYPE

Name:	**Immovable**
Definition:	An immovable object: land and attached objects. A single area of land or more particularly a volume of space, under homogeneous real property rights and unique ownership. Remark: By unique ownership is meant that the ownership is held by one or several owners for the whole Immovable. By homogeneous property rights is meant that rights of ownership, leases and mortgages affect the whole Immovable. This does not apply to specific rights as servitudes that may only affect part of the Immovable.
Aliases (optional):	Dutch: vastgoedobject
Feature attribute name(s):	
Feature association name(s):	to ImmovableComplex
Feature operation names(s) (optional):	N.A.
Subtype of:	RegisterObject

FEATURE TYPE

Name:	**RegisterParcel**
Definition:	Parcel subject to Registration
Aliases (optional):	
Feature attribute name(s):	legalSize
parcelName	
Feature association name(s):	ServingParcel
Feature operation names(s) (optional):	N.A.
Subtype of:	RegisterObject and Parcel

FEATURE ATTRIBUTE

Name:	legalSize
Definition:	The area of the parcel as described in legal (source) documents. This area can have been determined earlier in time and in general this area is not equal to calculated area from the spatial cadastral boundary vertices.
Value data type:	numeric
Value measurement:	square meters (or alternative from legal document)
Value domain type:	real; other data types (Area: ha.are.ca or integer: m²) as they can be in local use can be derived from this
Value domain:	positive real

FEATURE ATTRIBUTE

Name:	parcelName
Definition:	geographic name of the parcel as locally known
Value data type:	character
Value measurement:	from legal document
Value domain type:	
Value domain:	

FEATURE TYPE

Name:	ServingParcel
Definition:	Serves two or more RegisterParcels and is held in joint ownership by the owners of those RegisterParcels
Aliases (optional):	Dutch: mandeligheid; French: mitoyenneté
Feature attribute name(s):	
Feature association name(s):	RegisterParcel
Feature operation names(s) (optional):	N.A.
Subtype of:	Parcel

FEATURE TYPE

Name:	**Parcel**
Definition:	A single area of land, or more particularly a volume of space, under homogeneous real property rights (UN/ECE, 2004) or social tenure relationships. The whole domain is subdivided in

two types of regions (where it concerns the representation of real objects into the model): regions based on a partition (ServingParcel and RegisterParcel) and regions not based on a partition (NPRegion: non planar region; described within a NPRegion by TextParcel, PointParcel, or SpaghettiParcel's). Regions with a partition are completely covered by nonoverlapping parcels and can be represented in a topological structure (nodes, edges and faces and, depending on the dimension, solids). A Parcel may change its representation over time from TextParcel to PointParcel to SpaghettiParcel to RegisterParcel (fuzzy faces belonging to the same partition of space or surface).

Aliases (optional):	Dutch: perceel
Feature attribute name(s):	Urban, computedSize, dimension, spatialDescription
Feature association name(s):	AdminParcelSet
Feature operation names(s) (optional):	N.A.
Subtype of:	

FEATURE ATTRIBUTE

Name:	Urban
Definition:	Is Urban or Rural parcel (in case of Urban and Rural Cadastral system)
Value data type:	boolean
Value measurement:	from legal competence
Value domain type:	
Value domain:	

FEATURE ATTRIBUTE

Name:	computedSize
Definition:	calculated area based on the coordinates of the boundary points. This area is most of the time not exactly equal to the legalSize of registerParcel
Value data type:	real; other data types (Area: ha.are.ca or integer: m^2) as they can be in local use and can be derived from this
Value measurement:	spatial database
Value domain type:	
Value domain:	

FEATURE ATTRIBUTE

Name:	dimension
Definition:	dimension of Parcel: surface or volume
Value data type:	integer
Value measurement:	
Value domain type:	
Value domain:	2D, 3D

FEATURE ATTRIBUTE

Name:	spatialDescription
Definition:	spatialRepresentation
Value data type:	
Value measurement:	
Value domain type:	
Value domain:	ISO 19107

10 Metadata and Spatial Searching as Key Spatial Information Infrastructure Component
Future Standardization Developments

Marcel Reuvers and Henri J. G. L. Aalders

CONTENTS

During recent years a lot of attention has been focused on developing spatial information infrastructures (SIIs). Metadata is a major component for searching data by a SII. In this chapter the broader context of metadata is addressed, and open points for further work are also mentioned. Because of the background of the authors, this chapter has a standardization viewpoint.*

10.1 WHY CREATE METADATA?

Any organization providing information should disseminate data in a way that non-specialized users can discover, evaluate and use. The basic strategy is searching

---

* Material is reused from the work that has been done by the INSPIRE Drafting Team Metadata (Reuvers is chair of INSPIRE Drafting Team Metadata) and Dutch architecture papers.

for words or phrases in the contents resources as a hit-or-miss strategy. When the spatial resource of interest has some word or phrase uniquely associated with it, this can be quite successful but mostly, hundreds or thousands of irrelevant "hits" may be returned, as anyone can confirm who has spent frustrating hours searching for something whose name is a common word.

An alternative is to use metadata to describe resources in terms of certain well-defined attributes, such as resource title, geographic extent of the resource, resource topic category, or keywords. This allows users to search for keywords, names and phrases in particular contexts or a structured search. This means that effective use of spatial metadata is based on three components:

- A set of commonly understood terms that are used to describe the content of the information resources
- A standard grammar for connecting those terms into meaningful meta-data concepts
- A framework that allows the transfer and recombination of those metadata concepts across different applications and subjects

Together these three elements provide the architecture for spatial information description that can work across all associated subject areas (Aalders and Hunter 2002). Therefore, the mission for spatial data inventory applications is to make it easier to find resources using the Internet through the following functions:

- Developing spatial metadata standards for information search and retrieval
- Defining frameworks for the interoperability of spatial metadata sets
- Facilitating the development of community- and/or subject-specific meta-data sets that work within these frameworks

This may be done by providing user application software with tools as buttons, menus or navigational structures on a Web site (or, in the ultimate situation) by providing free-text search capabilities.

For example, an organization's name might be defined as a responsible party or as the distributor of a spatial data set, in contrast to having no such information or being one of the many organizations that are described in a document linked to the spatial data set. If this capability is combined with the use of "controlled vocabularies" (i.e. standardized lists of terms, such as abbreviations for countries or code lists for categories) and standardized formats for values such as time, dates, or longitude/latitude, it can greatly improve the efficiency of discovery. Also, if all spatial resources are assigned metadata such as a resource topic category, it becomes much easier for a user to find resources that match a query for a specific topic.

From the perspective of a governmental organization, it is important to help users obtain accurate and appropriate information: if users suffer some kind of loss as a result of finding incorrect or inappropriate information, they will make wrong decisions. In order to improve the discovery, evaluation, and application of government information, the metadata created to describe resources at different Web sites and by different organizations must share a common form and meaning, so that users do not have to learn a different set of terms and search strategies for each site they visit. Such "interoperability" is especially important for users who need to combine or compare information from multiple resources, but it is useful for any user attempting to discover information provided by government. This means that metadata standardization is needed to get better results in discovery and understanding.

10.2   STANDARDS-BASED APPROACH CHOSEN FOR INSPIRE

Many factors encourage the adoption of standards. Very detailed information on this topic is provided in the *GSDI Cookbook* (*GSDI Cookbook* 2004), but the following information is more particularly applied in the INSPIRE (Infrastructure for Spatial Information in Europe) context:

1. The importance of using a dedicated spatial metadata standard to support the implementation of a Global Spatial Data Infrastructure has been demonstrated by the different initiatives conducted since the early 1990s, particularly in:
   - The United States, with the development of the Content Standard for Geospatial Metadata by the U.S. Federal Geographic Data Committee (FGDC) and the presidential executive order that all federal government agencies were required to produce metadata for their spatial data holdings (ISO 2002)
   - Europe, with the experimentation of ENV 12657 (CEN 2005) and more recently with the emergence of national and subnational spatial data infrastructures and their increasing adoption of ISO 19115 (ENV/Temporary European Norm, CEN/Comité Européenne de Normalisation, ISO/International Organisation for Standardisation)
2. More than the lack of metadata, the lack of compatibility between the existing and upcoming metadata solutions is certainly one of the greatest challenges of the INSPIRE Directive (EU 2007, European Union). At this stage and due to the importance of the community, the use of a standard lexicon is a key to success.
3. The standardization activity in this geo-world has reached a level of maturity. A metadata standard dedicated to geographic information is available with the publication of ISO 19115. Its applicability to the European context was established with its adoption by CEN in 2005. The reference materials provided by the INSPIRE community for the establishment of these Implementation Rules (IRs) show a general endorsement of this international standard by the different European actors of the geographic information domain. Most of the legally mandated organizations and the spatial data information community (SDIC) have already adopted ISO 19115 or have ongoing activities to adopt it, as indicated by the results of the INSPIRE survey conducted in spring 2006 (INSPIRE 2006).

Access to an online metadata repository is fostered by standard interface specifications, such as the OGC CSW 2.0 (Geospatial Consortium Catalog Services for the Web) that can accommodate the use of different abstract metadata standards and related encoding, such as

- ISO 19115 and ISO 19119 through their ISO TS 19139 XML Schema implementation (CSW2 AP ISO)
- Dublin Core and its XML Implementation, which are relevant to ensure the relationship with other communities (XML/eXtensible Markup Language)

One should realize that multilingualism becomes a very important aspect because the Web connects a diversity of linguistic and cultural aspects from all over the world. So, the Web will fail to achieve its potential as a global information system, unless resources can be made available to users in their native languages, in appropriate character sets and with metadata appropriate to resource management. Here *internationalization* and *localization* become apparent, although they may be contradictory; whereas global resource discovery is best served by internationalization (using common conventions of practice, languages, and character sets throughout the world), the needs of any given community may be better served by supporting local conventions. Basic to the promotion of a global metadata architecture is to translate relevant specification and standard documents into a variety of languages. DCMI (Dublin Core Metadata Initiative) maintains a list of translations of its basic documents, as the European workshop on Learning technologies is maintaining translations of the LOM (Learning Object Metadata) specifications (Moellering et al. 2005).

The Dublin Core metadata element set (or the basic interlocking brick) is intended to support cross-subject search and retrieval. It can be thought of as a simplistic or pidgin metadata language that helps the user navigate through disparate subjects, languages and cultures. Adoption of the Dublin Core by governments, libraries, museums, archives, publishers, environmental science repositories, print and e-print archives, to name a few, testifies to its success in this role. There are emerging applications in the commercial sector as well, with health care organizations and financial industries using the Dublin Core as the basis for organizing and exchanging information.

10.3   METADATA IN A FEW WORDS

ISO 19115 defines metadata as "data about data."* This basic definition implies an unlimited scope to what can be seen as metadata. It allows some experts to see information as data or metadata with an unrealistic border between both, and also including data services in metadata. The INSPIRE Directive clarifies the definition of metadata as information describing spatial resources, making it possible to discover, inventory, and use them.

---

* INSPIRE has changed this definition to "information describing spatial resources, making it possible to discover, inventory and use them." This definition of metadata originates from the directive. It is compatible with the general definition of metadata provided in ISO 19115 and the OGC abstract specification for metadata: "data about data." It clarifies the expected role of metadata within the INSPIRE Infrastructure.

The metadata for those resources comprise

- Identification information, that is, information to uniquely identify the resource, such as
  - Title, abstract, reference dates, version, purpose, responsible parties
  - Geographic extent
  - Browse graphics (overview, thumbnail)
  - Possible usage
- Legal and security constraints
- Content description, that is, information identifying the feature catalogue(s) used and/or information about the coverage content
- Reference system information, that is, identification of the spatial and temporal system(s) used in the resource data
- Spatial representation, that is, information concerning the mechanisms used to represent the resource data spatially
- Quality and validity information, that is, a general assessment of the quality of the resource data, including
  - Quality measures related to the geometric, temporal, and semantic accuracy; the completeness or the logical consistency of the data
  - Lineage information, including the description of the sources and processes applied to the sources
  - Validity information related to the range of space and time pertinent to the data; to whether the data have been checked to a measurement or performance standard or to what extent the data are fit for the purpose

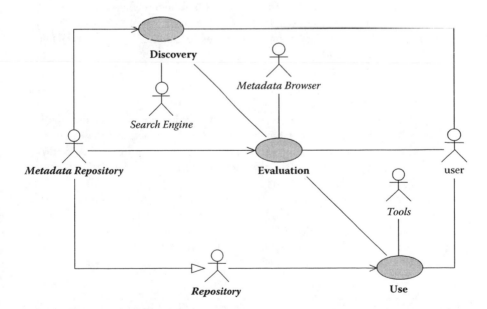

FIGURE 10.1 The INSPIRE Metadata Use Case.

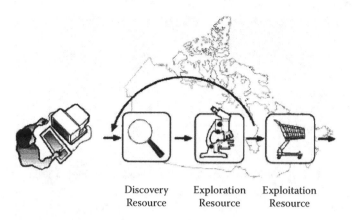

Discovery   Exploration   Exploitation
Resource   Resource   Resource

FIGURE 10.2   The three steps: discovery, exploration, and exploitation of resources.

- Portrayal information, that is, information identifying the portrayal catalogue used
- Distribution information, that is, information about the distributor of and options for obtaining the resource
- Maintenance information, that is, information about the scope and frequency of updating of the resource data

The INSPIRE Metadata Use Case (see figure 10.1) creates a general context federating the existing and upcoming metadata-based solutions around three base user activities* (see figure 10.2):

1. The *discovery* of resources. The user expects to identify a set of resources satisfying a basic set of search criteria. The user interacts with a search engine connected to a set of metadata repositories that document available resources. The search engine transfers the search criteria to the metadata repository and expects a minimum set of metadata related to the matching resources. It consolidates the answers and provides them to the user through an adapted interface.
2. The *evaluation* of available resources. The user has now identified a candidate resource, potentially as a result of the discovery activity, and wants to determine whether this resource satisfies his or her requirements. For this purpose it may use a metadata browser to examine more detailed metadata about the resource.
3. The *use* of adequate resources. The user has chosen a resource and some access and use rights have been granted to him or her. The resource is accessible and can be used through a series of dedicated tools. Metadata will support the user in fully understanding the data and using them properly, resulting in more reliable analysis and more confidence in the results.

---

* Some INSPIRE terms differ from the ones in the *GSDI Cookbook*, for example, "evaluation," "use" (INSPIRE) versus "exploration," "exploitation" (*GSDI Cookbook*).

TABLE 10.1
Organizational Benefits of Metadata

Data Archive	Data are the most expensive components of a GIS. Metadata is a means of preserving the value of data investments. This is of particular significance to local and regional governments experiencing rapid staff changes.
Data Assessment	GIS data development has shifted from data producers to data consumers. From a consumer perspective, metadata is the truth in labeling required to assess available data products. From the producer's perspective, metadata is a means of declaring data limitations and serves as a form of liability insurance.
Data Management	Metadata enables organizations to retrieve in-house data resources by specific criteria for global edits and annual updates.
Data Discovery	Metadata is the primary means of locating available spatial data resources via the Internet. Metadata is a primary public information resource as it is a nontechnical means of presenting technical information.
Data Transfer	Metadata is increasingly used by software systems as a means of properly ingesting data and by analysts as a means of properly displaying data.
Data Distribution	By building metadata in compliance with national standards, one can participate in the Global Spatial Data Clearinghouse. Participation promotes your agency and frees staff from answering data inquiries.

Different kinds of users may be involved in the different activities; some experts may evaluate the resources while operators may use them. There are cases where the user may be a software program performing automating searches. Depending on the users, different types of search engines, metadata browsers and tools may be necessary. From this perspective, this use case does not constrain the software market and it even creates a general context fostering the emergence of new markets and the satisfaction of new user requirements.

10.4   THE METADATA IMPLEMENTATION

Metadata production within organizations has recently become professional. There are a lot of initiatives where metadata is built up. The way this is done has to be critically reviewed. An example is that the metadata production is done after the data production process. Consequences are that the metadata are difficult to catch and that means that quality and completeness are not guaranteed. The integration of the metadata production within the data production* causes problems explained by the following reasons (Wayne 2005), but is the way to go:

1. Metadata standards are too extensive and difficult to implement.
2. Metadata production requires time and other resources.
3. There are few tangible benefits and incentives to produce metadata.

The achievement is to solve these three problems. Organizational benefits are listed in table 10.1.

---

* At the end the distinction between metadata and data will disappear in the production process. Advanced organizations are thinking in processes and information (models) and not in metadata and data.

10.5   THE IMPORTANCE OF TEMPORAL EXTENT

Besides the spatial extent, the temporal extent is worth mentioning because of the importance for temporal queries. The temporal aspect is also important in several domains, for example, earth science (see chapter 5) and cadastral registration (see chapter 9) (van Oosterom 2002, 2006).

Some general user queries on temporal extent might be

- A travel company wishes to search for the January snow climatology for the Alps, for inclusion in a brochure. The full temporal coverage might be the summary statistics of snowfall and snow cover for the period 1971 to 2000 taken over all Januaries. Here the discovery metadata are specific for area and for temporal extent. The classification of temporal extent is an ordinal reference system of named months.
- A hydrologist is looking for winter rainfall climatology in a catchment area. Here the ordinal classification of winter may be more important than a specific geographical position, as the hydrologist will expect to generalize to the catchment from whichever specific locations are found.
- A geologist may wish to find surface beds of Tertiary minerals across western Europe. Again the ordinal temporal classification is initially more important than location.
- A marine investigator requires weather information for a specific period for an accident at sea, the location being only approximately known.
- An archeologist wishes to compare sites across regions that are known to be active in the same time period. The discovery is to find locations for a known period.
- Anyone who wants a weather forecast for three days hence. For a weather forecast, the sequence of weather is usually more important and may be more accurately forecast than the expected weather at a specific location and time.

Temporal aspects vary quite significantly with respect to the application domain. The semantics of the temporal extent will be in development for the next years. The domain standardization will be leading here. From our experience, INSPIRE will be an important catalyzer to achieve this.

10.6   THE CHANGED ROLE OF METADATA IN
SERVICE-ORIENTED ARCHITECTURE (SOA)

Any organization wanting to install an SOA application has to migrate its application architecture. Figure 10.3 describes a general application architecture for an individual organization.

An application architecture brings the functionality of applications in logical layers with standardized interfaces. This creates more flexibility and better possibilities for maintenance because changes in a logical layer do not have consequences for the other layers, as long as the interfaces between layers do not change.

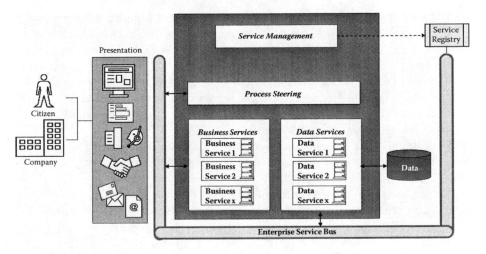

FIGURE 10.3   The Dutch Reference model service layers.

Within the Dutch Reference model for Architecture the following layers can be distinguished:

- Data
- Data services
- Business services
- Presentation
- Process steering
- Service management

The layer service bus is not worked out in more detail because its task is mainly logistic (way of delivery, envelope). In The Netherlands one applies a basic bus for interfaces and security and a richer bus with possibilities for schema translation, orchestration, subscriptions and so on.

Data: The data are all digital information that are maintained in databases, file systems, document management systems, geo-information systems, etc.

Data services: The data services give access to data. With these services data can be created, changed, deleted or viewed. Data services are mostly offered by a provider without knowing what the user wants, where a user can be a human or an application. A user has to change his environment in the way the data are served. In the geoworld examples of these services are Web mapping services (WMS) of Web feature services (WFS). The metadata described in the previous sections are metadata over data services, making use of ISO 19115, ISO 19119, ISO TS 19139 and OGC capabilities. This is the "classic" way of describing and using metadata.

Business services: Business services supplies services to users (humans or applications). A characteristic for business service is that the user knows beforehand which performance will be delivered but not how this is done. A business service fulfils that delivered information is focused on the process in which it is used. These business services may use data from the data services. A business service can make use of WFS Filter encoding but more regular and needed is the application of SOAP (Simple Object Access Protocol) and WSDL (Web Services Description Language) as a service description. The metadata for these services are not focused on the data but on fulfilling the business needed. That makes the description of these services different from the "classic" way.

Presentation: The presentation layer takes care of the interaction with the user. Information for applications services is presented in a way that fits the channel used.

Process steering: These services make use of other services. Service agreements are needed on the method of interaction with the user. This has consequences for the process implemented within organizations. Process steering takes care that the business services needed are chained in the right order. This is about orchestration, choreography, business-process management and workflow management. WS-BPEL (Web Services—Business Process Execution Language) is a schema language that is often applied in these processes. The metadata have to support these processes. Questions arise, such as, how are the metadata described for a new service as an outcome of a WS-BPEL process and what is the role of the metadata of the individual business services that are part of it?

Service management: An organization has to take care that services will fulfil in time the requirements of the users. That means creating new services and removing or changing existing services. For several reasons it is important to describe and publish the services in a registry. In this way the provider has to register the services only once instead of many times. Examples of these service registries are UDDI (Universal Description, Discovery, and Integration protocol), ebXML (Electronic Business using eXtensible Markup Language), RIM (Registry Information Model) and CSW (Catalogue Service). These registries also fulfill other requirements: the metadata in these different kind of registries are also different on aspects based on the purpose of the service itself. A good investigation of the differences among these registries is done by ISO TC/211 in the ad hoc group on the study of ebXML RIM, 2007-02-16, http://www.isotc211.org/protdoc/211n2165/.

In the geoworld we can distinguish three types of registries:

1. Discovery services—data set metadata, or basic information about a data set, for example, about the identification, extent, quality, feature types, spatial reference, and distribution.
2. Publishing services
   Service descriptions—basic information about a service, for example, a description of the operations and their parameters as well as information about the geographic information available from a service offering.

- List of service types (service taxonomy).
- Data specifications—detailed description of one or more data sets that will enable it to be created, supplied to, and used by another party.

3. Register services—the number of registers that need to be maintained in the infrastructure can be significant. As a result, a clear and sustainable operational model forms a key part of the setup of the infrastructure. A starting list of potential registers, that is, kinds of spatial information items, includes

- Feature catalogues, which are catalogues containing definitions and descriptions of the feature types, their attributes and associated components occurring in one or more data sets, together with any operation that may be applied as part of a data specification.
- Application schemas, conceptual schemas for data required by one or more applications. Part of a data specification and specified in a formal conceptual schema language (typically UML [Unified Modelling Language]).
- Code lists, a dictionary describing the attribute value domains for selected property types in a feature catalogue/application schema. In these cases the value domain is not fixed in the feature catalogue/application schema, but is managed separately; that is, this establishes a controlled vocabulary.
- Thesauri, similar to code lists with additional information on how terms of the vocabulary relate to each other (hierarchies, etc.). It is unclear whether this is needed in the first step.
- Coordinate reference systems, a dictionary of coordinate reference systems, data, coordinate systems, and coordinate operations that are used in data sets.
- Units of measurements, a dictionary of units of measurement that are used in data sets.
- Spatial object identifier namespaces—a mechanism is required to guarantee the uniqueness of feature identifiers across various content providers. One approach is to use existing "local" identifiers of the provider, but define namespaces to distinguish between different providers (and between different offerings of a provider). These namespaces need to be managed.
- Portrayal rules, or rules that are applied to a feature to determine the portrayal of a feature in a map.
- Symbols, depictions to be used in portrayal rules to describe the styling of features in a map.

As we can see many services will exist; the context is much broader than only a registry for metadata in the "classic" way. Besides the discussion on which registries are needed on the software, the interesting part also will be which metadata will be needed to fulfil the user requirements. It is expected that more and more metadata will be interpreted by machines instead of humans.

10.7   THE GEOSPATIAL WEB

The Geospatial Web is growing fast: the most significant platforms are Google Earth, Microsoft Virtual Earth and NASA (National Aeronautics and Space Administration) World Wind. Behind this there is new technology growing, such as geo-tagging, the Semantic Web, open source, open data, and much more, mostly driven by the general Web 2.0 developments. Scharl and Tochtermann (2004) provide an excellent description of new developments regarding the Web.

The metadata way of Web 2.0 thinking comes from a different point of view than the current metadata thinking. The metadata way of Web 2.0 thinking concentrates on, for example:

- Conceptual search by ontologies (making use of OWL, the Web Ontology Language), chapter 5, Conceptual Search: Incorporating Geospatial Data into Semantic Queries and Chapter 23, SWING—A Semantic Framework for Geospatial Services)
- Making use of communities for tagging geodata and building metadata
- Creating geo-tagged pictures by address (geographic extent), chapter 15, Sharing, Discovering, and Browsing Geo-tagged Pictures on the World Wide Web

and much more. This means that the processes of building metadata can be enriched by making use of the Web 2.0 developments. Otherwise it is important to align the current metadata developments with the Web 2.0 developments, especially when the geocommunity wants to discover not only geographic data but also new data (as pictures and others sensors) that have a location too.

The Geospatial Web will have an impact on the current metadata in building the metadata and conceptual search. New types of metadata for different kinds of data will arise. Therefore it is important to align continuously the current metadata developments with Web 2.0 developments. To arrange this, sustainable standards are needed and the work of ISO TC/211 and OGC with the Semantic Web world of W3C needs programming and more detailed cooperation. For example, the hierarchical structure of ISO 19115 makes this alignment more difficult.

10.8   GENERAL CONCLUSION

Metadata will support the user in fully understanding the data. It appears that metadata for data and data services are well arranged in the field of standardization. On the other hand, we cannot sit back because new challenges come up and the traditional background of data (maps) focusing will not help us. These new challenges are pointed out in sections 10.5, 10.6, and 10.7. Of course there are and will be other important developments, but these will come from different directions and from different communities. This means that the next step in metadata development will require more and more cooperation with other domains.

REFERENCES

Aalders, H. J. G. L., and G. J. Hunter. 2002. *Spatial Data Standards.* UNESCO-EOLSS Encyclopedia Knowledge for Sustainable Development: An Insight into the Encyclopedia of Life Support Systems, Paragraph 6.72.3.4. August.

CEN. 2005. *EN ISO 19115 Geographic Information—Metadata.* EN European Norm.

ENV. 12657: 1998, Geographic Information—Data Description—Metadata.

EU. 2007. *Directive of the European Parliament and of the Council establishing an Infrastructure for Spatial Information in the European Community (INSPIRE),* January 2007.

GSDI Cookbook. 2004. http://www.gsdi.org/docs2004/Cookbook/CookbookV2.0.pdf.

INSPIRE. 2006a. Drafting Team Metadata, Draft Implementing Rule Metadata as Provided to the SDICs and LMOs, see http://www.ec-gis.org/inspire/reports/ImplementingRules/draftINSPIREMetadataIRv2_20070202.pdf.

INSPIRE. 2006b. see http://www.ec-gis.org/inspire/reports/INSPIRE_Metadata_Survey_2006_final.pdf.

ISO. 2002. *ISO 19115: 2002 Geographic Information—Metadata.* Geneva: International Organization for Standardization.

ISO. 2003. *ISO 19119 Geographic Information—Services.* Geneva: International Organization for Standardization.

ISO. 2004. *ISO TS 19139 Geographic Information—Metadata—XML schema implementation.* Geneva: International Organization for Standardization.

Moellering, H., H. J. G. L. Aalders, and A. Crane. 2005. *World Spatial Metadata Standards.* Amsterdam: Elsevier (published on behalf of the International Cartographic Association).

Onsrud, H. *GSDI Cookbook v2.0* (PDF) January 2004.

Scharl, A., and K. Tochtermann. 2004. *The Geospatial Web, How Geobrowsers, 2004 Social Software and the Web 2.0 Are Shaping the Network Society.* Berlin: Springer.

van Oosterom, P. J. M., and C. H. J. Lemmen. 2006. Spatial Data Management on a Very Large Cadastral Database. *Computers, Environment and Urban Systems* 25 (4/5):509–528.

van Oosterom, P. J. M., B. Maessen, and C. W. Quak. 2006. Generic Query Tool for Spatio-Temporal Data. *International Journal of Geographical Information Science* 16 (8):713—748.

Wayne, H. 2005. see http://www.fgdc.gov/metadata/documents/InstitutionalizeMeta_Nov2005.doc/view.

11 The Spatial Information Infrastructure as Part of the Information Infrastructure

Bo Overgaard and Thorben Hansen

CONTENTS

The Danish society has a long tradition of maintaining national registers supporting government related to citizens, business, land, environment, etc., and today these registers are available as central databases. Since the 1990s the basic national map resources have been converted into digital spatial representation of real-world objects. With spatial objects sharing keys with the traditional national registers, whole new perspectives emerge in dealing with spatial information as "just" an additional dimension to traditional register information.

E-government initiatives put focus on utilizing existing digital information actively in digitally supported workflows. Combining traditional register information with spatial information offers new ways of supporting IT solutions, with focus on intuitive, user-friendly user interfaces and support of well-informed decisions.

Opening existing digital information to broader use requires a new approach to how information is shared and how it can be embedded into e-government solutions. A business model with basic spatial and nonspatial information as infrastructure elements must be developed. Service-oriented architecture, technology and data standards, agreements on sharing information, and a partnership model that supports the role of both government infrastructure providers and private system integrators are important facilitators.

165

This chapter describes the approach taken in Denmark from both a technology and a business perspective. Emphasis is put on the business aspects of defining the interface between the information infrastructure (II)/spatial information infrastructure (SII) and its use, establishing a platform for fruitful collaboration between private and public partners. Examples are given on actual implementations based on the approach.

First the reader will be introduced to the status of digital registers, both spatial and nonspatial, in Denmark. Afterwards the chapter describes how standards and changes in business models have been a driver for the use of Web services in Denmark. To exemplify the advance the chapter ends with some cases describing how spatial information infrastructure is part of the general information infrastructure.

11.1   DIGITAL REGISTERS

Managing a modern society requires substantial information about the assets of the society. In highly regulated societies like the Danish, government needs detailed information about fixed assets like land and buildings and how they relate to citizens and businesses. The need is related to multiple purposes, such as securing ownership, taxation, physical planning, and emergency response.

In the late 1960s and in the 1970s, a number of national registers were built containing such information. In line with the technological capabilities of the time, the registers were built with simple tabular attribute information.

Some of the most interesting and most applied of these registers include:

- Real property valuation assessment and land title registration based on cadastral parcel identification (ref: www.kms.dk, www.domstol.dk, www.skat.dk)
- Centralized civil registration (CRS, established in 1968) containing basic information about all citizens, including their addresses given as street code and house number (ref: www.cpr.dk)
- Building and dwelling registration (BDR, established in 1976) containing technical information about buildings and dwellings, such as areas, numbers of floors, use, construction materials, installations, etc., and including addresses given as street code and house number (ref: www.bbr.dk)

Whereas land parcel identification was already a well-known key to uniquely identifying land, the street code and house number address system was established as a unique key with these systems. Since the 1970s, many more registers have been established. Today the address system is used extensively in other systems; however, often with modifications.

Traditionally registers were built for specific purposes and part of monolithic solutions for different authorities. Data were typically not accessible for other authorities or for the public.

An important agreement was made in 2002 between the state and the municipalities. As a result of this agreement, a common server was established in 2002, called OIS (Public Server of Information) (ref: www.ois.dk). The server contains replicas of a number of the (nonspatial) basis registers that are related to real estate and come from different government sources. Citizens can, via a Web site, query information about their own property and some information about property owned by others. Data for professional use are distributed via private licensing agents (companies such as Grontmij | Carl Bro). Some licensing agents have chosen to provide Web services so the data can be used in other applications; others just provide a database dump.

As an alternative to going via private licensing agents, a number of the government organizations responsible for data are offering the information as Web services directly from the data source. In 2007 a new version of the building and dwelling register (BDR) was released. The new version will offer direct Web service access to data.

11.2   DIGITAL GEODATA

Digital methods were introduced in the mapping environment in the 1980s and 1990s. The initial focus was on improving production methods. At the same time, there has been a growing understanding that a map is not only an image representation of the real world, but that the elements in the map are spatial representations of real-world objects. The full potential of digital maps can only be released if the "mapping" information is seen as another dimension added to real-world objects—objects that to some extent are already described in existing digital registers.

Major emphasis has therefore been put on breaking the map into its underlying objects and to add key information that links each object with its description in other registers. "Geo-keys" that serve as unique object identifiers in both the spatial and the nonspatial environment allow for more informed use of existing government registers by adding a spatial dimension, and new knowledge can be gained by spatially correlating objects that are not otherwise correlated.

By taking this approach, the focus moves to providing geodata as a spatial information infrastructure that can add a spatial dimension to other registers and can provide a platform for uncovering spatial interdependencies that cannot otherwise be uncovered. Maps are merely the presentation layer that can be added to the geodata to provide an intuitive representation that communicates well to most users.

The cadastral parcel map is one of the most significant data sets when it comes to adding a spatial dimension to multiple existing registers. It was turned digital in the late 1980s and early 1990s, providing a spatial dimension to all registers containing parcel or real estate information. A street map with spatial representations of all streets with street codes and an address register with coordinates to all addresses with street code and house number were available at the turn of the century.

Since 2001 the National Survey and Cadastre has provided some of the basis geodata registers as Web services, to start with as WMS and later on also as WFS via the so-called Digital Map Supply. Using geo-keys, information from other registers can be geo-coded via the Digital Map Supply. The geo-keys are today based on the SOAP protocol.

11.3   E-GOVERNMENT

First generation digitization in the government sector focused on building digital registers that were maintained via traditional government procedures, typically in monolithic environments, where updates are initiated by manual workflows and input by the responsible authority, and where the contents of the registers do not flow easily to other authorities where the information is relevant.

With the advent of the Internet and the general progress in technology, the focus is now on digitally supported workflows, where updates are fed directly from the initiating source, and where information is available on the Internet on an as-needed basis. The role of the authorities in this paradigm is to support relevant workflows on the Internet; to authorize changes based on relevant digital government acts and to define the rules for how, to whom and under which circumstances information can be shared.

This change in paradigm is tagged e-government. In 2001 the Ministry of Finance formed the Digital Task Force (ref: www.e.gov.dk) project to coordinate and push the change. The project is headed by the Ministry of Finance and has a steering group with representatives from state and local government.

So-called service communities are formed with reference to the Digital Task Force. Service communities are cross-government collaborations between authorities with a need for coordination in order to facilitate e-government. One such service community was formed in 2002 for geodata. It is headed by the Ministry of Environment and community members include the Ministry of Economic and Business Affairs; Ministry of Transport and Energy; Ministry of Food, Agriculture and Fisheries; Ministry of Science, Technology and Innovation; Association of Municipal Governments; and Association of Regional Governments. The Service Community for Geodata (ref: www.xyz-geodata.dk) takes initiatives to form committees for collaboration and gives recommendations in support of establishing a spatial information infrastructure.

Another major e-government initiative is the so-called Public Information Online (ref: www.oio.dk) initiative. It is headed by the Ministry of Science, Technology and Innovation with reference to the Digital Task Force. The initiative gives recommendations on communication and information technology standards and drives standards for data; it shares information about the initiatives and activities concerning technology aspects of e-government and e-solutions to citizens and enterprises. Recommendations are given within the areas of communication, security, IT-architecture, standards, availability, management, data exchange, and shared solutions.

The Digital Task Force initiative is aimed at e-government in Denmark. However, it is important to also see the local requirements in an international context. With the INSPIRE directive, the European Union has defined the framework of a European geospatial infrastructure that all member states must adhere to. The principles of the e-government initiatives and the INSPIRE directive are very much supporting each other, and implementation of the INSPIRE directive will further boost the development by creating a comprehensive spatial information infrastructure that can also be used across the borders of the EU countries.

11.4 STANDARDS

As mentioned above, the focus is on e-government and standards, and the Ministry of Science, Technology and Innovation is chairing the work with reference to the Digital Task Force. In 2003 a "white book on IT-architecture" was published (ref: www.oio.dk/arkitektur/soa). The white book emphasizes the benefit of service-oriented architecture (SOA). Later the same year the Public Information Online Web site (www.oio.dk) was launched. The Web site offers specialized texts such as manuals, guidelines, presentations, and reports on a wide range of subjects. Among these are enterprise architecture, metadata, software strategies, standard contracts, digital signatures, usability, XML, IT security, and benchmarking. Among the projects is the Danish OIOXML project (www.oio.dk/dataudveksling/danishXMLproject) and the Danish e-Government Interoperability Framework (ref: http://standarder. oio.dk/English/). The Interoperability Framework contains more than 600 selected standards, specifications, and technologies used in e-government solutions, divided into three categories: Process Standards, Technical Standards, and Data Standards. Among the standards approved are WMS and GML specified by the Open Geospatial Consortium (OGC). The geodata interest organization, Geoforum, has published guidebooks about WMS, WFS and GML, supported financially by the geodata service community. All together this has had a major impact on the use of the acceptance of the OGC specifications in Denmark. Some of the national distributors of standard GIS applications can report that the high awareness of OGC specifications in Denmark has put pressure on international software companies to make their products OGC compliant.

The goal is that the Danish OIOXML project will

- Improve the exchange of data both internally in the public sector and between the public and private sectors.
- Improve the processing of data, and make easy access to already collected data and the reuse of these.
- Make it easier to implement e-services.

To fulfill this goal, two initiatives were started in 2001, the establishment of the so-called Infostructurebase and the standardizing work. The Infostructurebase contains information about the content of government databases and how to get access to these. A project tender provided the background for the implementation of the Infostructurebase. The standardizing work determines standards for exchange of data between the public authorities and between public and private institutions (ref: http://isb.oio.dk/info).

11.5   CHANGE IN BUSINESS MODEL

As mentioned earlier, e-government is a change in paradigm, where workflows are supported digitally, updates are fed directly from the initiating source, and information is available on the Internet on an as-needed basis. This change in paradigm must be supported by a change in business model that from an overall perspective allows the players to play their role the most suitable way.

The players involved are (1) the commissioner responsible for a specific e-government solution, (2) the users of the solution, (3) the providers of common Web services used by multiple solutions, and (4) the system integrator that builds the specific

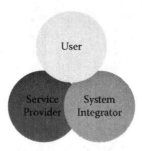

FIGURE 11.1   The relationship between user, service provider and system integrator.

solution. Compared to the old monolithic solutions, the role of the service providers is new, and special attention must be paid to how these new players fit in so that the services provided are easy to embed and add maximum value to the solution.

The Web services can be seen as an infrastructure. The interface between the infrastructure and the users of the infrastructure must be well defined, from both a technical and a business point of view.

In 2001 the National Survey and Cadastre changed its business focus to be the provider of a spatial information infrastructure. This change had a major impact on the business model and system development in Denmark. When the Digital Map Supply was launched in 2001, the declared goal was to offer an SOA-based spatial information infrastructure for cadastral, topographical and nautical information dealing with both the technical and the business aspects of such an infrastructure.

The main elements of the Digital Map Supply are

- Defining technical standards for the Web service interface
- Defining the service level for the Web services
- Defining the conditions for use of common Web services (agreements on sharing)
- Establishing a program encouraging system integrators to use the Web services

The conscious focus on all of these aspects in one approach has been a major contributor to the success of the Digital Map Supply. Today more than 200 organizations can include Web services from the Digital Map Supply in their solutions, more than 20 system integrators are attached to the developer program, and each month the Digital Map Supply serves more than 5 million requests.

Today more governmental organizations are offering information that they are responsible for as shared Web services for embedding into e-government solutions. The trend is that more such Web services are being established, in both the spatial and nonspatial areas.

11.6   ACTUAL DEVELOPMENTS: SAMPLE CASES

In the summer of 2004 a political settlement that changed the regional and municipal structure was reached. The reform itself took effect January 1, 2007. Prior to this date more than 270 municipalities and 14 counties existed in Denmark. The reform reduced the number of municipalities to 98, the 14 counties were abolished, and 5 new regions were founded instead.

In many ways the counties had been at the cutting edge of the use of GIS in Denmark. With the counties being abolished more or less all the assignments that involved the application of GIS were divided between the state and the new (larger) municipalities. Only few spatial data are today maintained by the new regions.

In preparation for the reform, the Association of Municipal Governments carried out some field studies in 2005 on how to migrate the tasks undertaken by the counties to the municipalities. The studies showed that the counties used GIS as an important tool in their daily work with environmental issues, and that the link between GIS and the document management systems containing government acts relating to land is essential for the administration. In 2006 a guide was made for the municipalities stressing the need for GIS when taking over the tasks.

Before the reform, the counties were responsible for almost all of the environment-related administration in Denmark. After the reform, the responsibility was split between the municipalities and the Ministry of Environment. As a consequence of this change, the data from the former counties were brought together in a number of shared systems called Danmarks Miljøportal (the Danish Environment Portal). The administration of environmental issues faced an important challenge that had to be dealt with when the reform became effective: some data will be maintained by the state and some by the municipalities, while the state, the regions, and the municipalities must be able to use the data across the administrative borders. Therefore, the only right solution was to establish these shared systems.

11.6.1 DANISH AREA INFORMATION SYSTEM

One of the major subsystems of the Danish Environment Portal is the Danish Area Information System (http://kort.arealinfo.dk/); see figure 11.2. This system contains more than 30 national themes plus a substantial number of local themes covering each of the former counties. All together there are over 1,000 data sets in the system. DAI is divided into a data production system and a presentation system.

In the presentation system all data can be viewed and queried in a Web-based GIS. Moreover, there are also WMS and WFS access options, which means that data also can be viewed on local desktops or Web GIS systems together with local data sets. The data are public and free of charge, and as a consequence of this, municipalities can design their GIS setups in such a way that their employees will always use the latest version of data in their daily work, without having to download and convert data into their own local databases. However, it is also possible to download data in GML and other formats like ESRI shape and MapInfo tab. The Web-based GIS uses WMS/WFS and SOAP services from the Digital Map Supply as base maps and for address searching, etc. Figure 11.3 gives an example of an address search dialog based on data from the SOAP geo-keys.

In the production system there are two ways of maintaining data. Data can be locked and downloaded for offline editing. If the municipality chooses, it can use its own desktop GIS for editing. When the editing is ended, the data can be uploaded to the

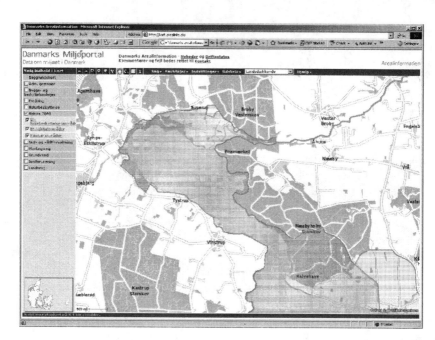

FIGURE 11.2   (See color insert.) The Danish Area Information System showing Natura 2000 protection areas.

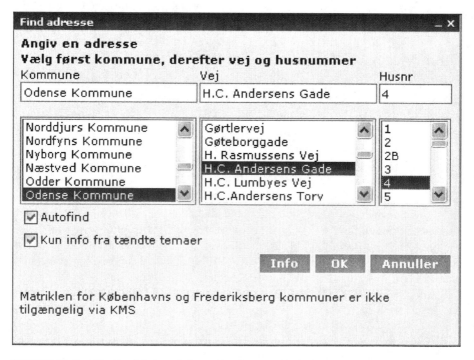

FIGURE 11.3   The Danish Area Information Systems find address dialog. Data are supplied by the SOAP geo-key services from the Digital Map Supply.

system for everybody else to use. The other possibility is to use the online editing tool. This is a Java applet-based Web digitizing tool that can be used by the municipalities without any cost. The Web-based digitizing tool works directly on the data in the editing system. The tool uses the WMS and WFS from the system and from other sources like cadastral maps from the Digital Map Supply. The editing tool is quite advanced: it combines WMS and WFS with possibilities to snap, create buffers, and build new objects by combining objects from other layers including the online cadastral map.

The only way data can enter the system is through a SOAP Web service. The online editing tool uses the SAOP service directly, but if data come from offline editing, there is a pre-process that converts the data to GML and then sends them to the SOAP service for updating.

The system use a SOAP interface and not the transaction WFS, known as WFS-T. Because the system has some business logic implemented so users can only edit data in their own area and rules about overlapping, polygons crossing borders, etc., the standard WFS protocol was not sufficient. But basically the SOAP services have much in common with WFS-T, including the use of GML.

The Danish Environment Portal, including the Danish Area Information System, is owned by the Danish Ministry of the Environment, the Association of Municipal Governments, the Association of Regional Governments and the Digital Task Force, in partnership. In that sense the organization is quite unique.

11.6.2 PlansystemDK

Since 1989 a register has existed containing physical planning information. From the beginning the system was not designed to handle the spatial dimension. In

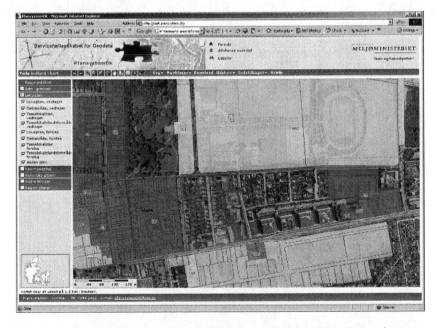

FIGURE 11.4   (See color insert.) PlansystemDK showing local physical planning areas.

2006 the system was replaced by a new register called PlansystemDK (http://www. plansystem.dk/); see figure 11.4. This register is able to handle spatial information. PlansystemDK is a national register and is based on service-oriented architecture. The system is unique in the sense that the municipalities are responsible for the content and validity of the data, whereas the state (the Ministry of the Environment, Danish Forest and Nature Agency) is responsible for the shared register and the system.

Like the Danish Area Information System, the only way to report data into the system is through a SOAP Web service that accepts GML. And likewise it is possible to do both online and offline editing of data. The PlansystemDK uses base maps from the Digital Map Supply and was the first system in production that used the new WFS capabilities from the Digital Map Supply. And naturally all the data in the system are reachable are WMS as well as WFS (figures 11.5 and 11.6).

The establishment of PlansystemDK is related to a reorganization of the registration of (district) plans in Denmark, which will take effect in 2008. From that time the district plans must no longer be registered in the land registry in order to obtain legal effect; registration in PlansystemDK will secure their legal validity.

In parallel with the actual deployments as illustrated above, SII is also continuously being improved and expanded.

The Central Business Register (CVR) contains primary data on all businesses in Denmark and covers both public and private businesses. In addition, the CVR contains detailed information on all limited companies, including fiscal reports, management and financial information, status, etc.

Private companies can buy these data, and the public sector can use the data free of charge. At the Web site www.cvr.dk there is a front end where different search opportunities make it possible to look at the data in the CVR. However, the CVR data are also accessible through download or they can be integrated in other systems by using a new online SOAP Web service. There is no spatial information in the CVR, but data can easily be geo-coded using the postal address.

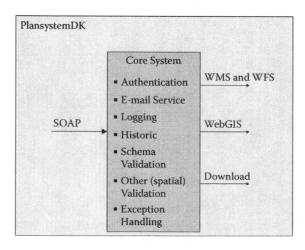

FIGURE 11.5   PlansystemDK: the core system.

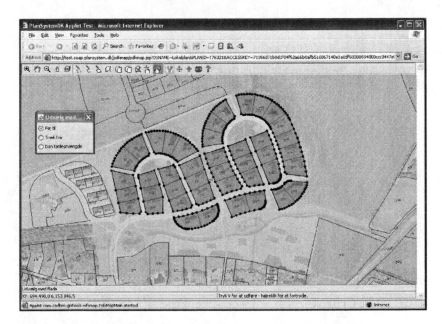

FIGURE 11.6 (See color insert.) PlansystemDK online editing tool, which provides advanced digitizing using WFS directly in a browser.

The Digital Map Supply has offered basic address geocoding Web services for years. The public Address Web-Services (AWS) is a new project still in tender, aimed to be realized in 2008. The objective is to provide a comprehensive set of SOAP-based Web services to the general IT developers so the use of address information in both e-government and commercial IT solutions is strengthened. AWS will consist of several services, including phonetic methods, etc. The aim is that common IT solutions will use the official way of spelling street names, use the right postal districts and ensure that addresses used actually exist.

AWS will be free of charge for both public and private companies and data will include spatial information (coordinated address points). The AWS project is established in cooperation between the National IT and Telecom Agency, the National Survey and Cadastre, and the Danish Enterprise and Construction Authority.

11.7 CONCLUSION

The practical implementation and application of a Danish SII has developed rapidly over the last few years. The main drivers for this development are

- Substantial existing government information in digital national registers
- E-government initiative driven by the Minister of Finance
- Recognition of the importance of "where" as a backbone for e-government
- Urgent need due to change in the Danish administrative structure

Instrumental for the achievement has been the use of service-oriented architecture and related standards in combination with a conscious focus on establishing a business model and a partner program supporting the cooperation between government bodies responsible for the (spatial) information infrastructure and private companies working as system integrators to embed the infrastructure into actual solutions.

Index